Teacher Certificati

Math High School

Contributing Authors
Patty White, BS
Heath Gatlin, BS
Jeff Limber, BS
Terresa Barton, Ph.D

Publisher
Sharon Wynne

Editors
David A. Collins
Terresa Barton, Ph.D
Patty Wynne, BS

Math Typist
Julie Carnevale, BS

To Order Additional Copies:
Xam, Inc.
99 Central St.
Worcester, MA 01605
Toll Free   1-800-301-4647
             Phone:1-508 363 0633
Email: winwin1111@aol.com
Web   www.xamonline.com
Fax: 1-508-363-0634

You will find:
- Study Guide – content review
- Test Prep - Sample Test

XAM, INC.
Building Better Teachers

Copyright © 2004 by XAM, Inc.

Note:
The opinions expressed in this publication should not be construed as representing the policy or position of the National Education Association, Educational Testing Service, or any State Department of Education.

All rights reserved. No part of this publication may be reproduced or transmitted in any form or by any means, electronic or mechanical, including photocopy, recording or any information storage or retrieval system, without permission in writing from the publisher, except where permitted by law or for the inclusion of brief quotations in the review.

Printed in the United States of America

New York: Mathematics High School
ISBN: 1-58197-328-4

TEACHER CERTIFICATION EXAM

INTRODUCTION

About this Study Guide

The XAM Math Secondary study guide is designed to review secondary school mathematics knowledge and competencies in preparation for teacher certification exams. XAM has in depth definitions and examples as a follow up to the actual test makers' introductions and pamphlets. The mathematics concepts are derived from the content on the teacher certification exam.

Content knowledge tests for math secondary covers the topics found in the table of contents. They may use some or all of the topics. Some exams require the use of a graphic calculator and others forbid it. You must check with your state certification agency to gain clarification. Praxis states in 2003 are requiring the use of graphic calculators.

The math secondary test is given in 46 states. Florida, New York, Texas, Oklahoma, Michigan, Massachusetts, and Colorado use state specific tests. This study guide lists and explains most of the topics on most of the tests. Praxis is a test given in the remaining states. Test at a Glance lists numerous formats, including Math: Content Knowledge (0061) and Math: Proofs, Models, and Problems Part I (0063) and Part II (0064). Details of Test at a Glance can be found at www.ets.org.

Some states such as Florida and Georgia have optional divisions of Math 5-9 and Math 6-12, which overlap to some degree but essentially refer to middle school and high school math. Xam makes additional math guides exclusively for middle school.

The Praxis exam contains 50 multiple-choice questions divided into three sections. Arithmetic and basic algebra, geometry, trigonometry, and analytical geometry form the first section, comprising approximately 34% of the exam. The second section consists of functions and their graphs and calculus. This section constitutes approximately 24% of the exam. The last section consists of approximately 42% of the questions on the exam. The competencies in this section include probability and statistics, discrete mathematics, linear algebra, computer science, and mathematical reasoning and modeling.

MATHEMATICS HIGH SCHOOL

TEACHER CERTIFICATION EXAM

TABLE OF CONTENTS

Competency — **Page**

1. Properties and Operations the Real Number System — 1
2. Knowledge of the Basic Properties of Set Theory — 6
3. Ratio and Proportion, Percent, Fraction, and Decimals — 7
4. Comprehension of Number Theory — 13
5. Geometric Concepts, Formulas, and Constructions — 20
6. Probability and Statistics — 28
7. Interpret and Extrapolate from Graphs, Tables and Charts — 33
8. Algebraic Function, Relations, and Their Graphs — 36
9. Comprehension of Integral Exponents — 49
10. Polynomials and Rational Algebraic Expressions — 51
11. Polynomial Factorization Over the Rational Numbers — 56
12. Mastery of Rational and Irrational Expression — 60
13. Knowledge of Quadratic Equations and Inequalities — 70
14. Proficiency in Evaluating and Graphing Functions — 81
15. Direct and Inverse Variations — 89
16. Ability to Solve Equations and Inequalities — 91
17. Polynomial Functions — 111
18. Solving Problems Involving Rational Exponents — 117
19. Performing Operations with Complex Numbers — 120
20. Solving Problems Involving Sequences and Series — 122
21. Solving Problems Involving Permutations and Combinations — 126
22. Knowledge of the Binomial Theorem — 127
23. Knowledge of the Fundamental Geometric Concepts — 129
24. Knowledge of Types of Polygons and Their Angles — 133
25. Ability to Investigate and Apply the Concepts of Congruence — 142
26. Ability to Apply the Properties that Relate to Similar Polygons — 148
27. Ability to Apply Relationships that Exist in Right Triangles — 150
28. Properties of Angles and Lines that are Appropriate to Circles — 156
29. Basic Constructions with a Compass and a Straight Edge — 164
30. Axiomatic Approach in Developing Proofs for Theorems — 173
31. Lines, Angles, Triangles, Quadrilaterals, and Circles in Proofs — 183
32. Euclidean and Non-Euclidean Geometries — 190
33. Perimeter and Area of Plane Figures and Surface Area and Volume of Regular Solid Figures — 191
34. Compute Descriptive Measures from Given Ungrouped Data — 201
35. Understanding of Matrix Algebra — 203
36. Circular/Trigonometric Functions and Their Inverses — 211
37. Ability to Prove Circular/Trigonometric Function Identities — 212
38. Proficiency in Graphing Trigonometric Functions — 215
39. Ability to Solve Problems Involving the Solution of Triangles — 216
40. Understanding of the Theory of Functions — 226
41. Knowledge of Circular/Trigonometric Functions — 229

42. Mastery in Performing Operations on Vectors	231
43. Determining Distances on a Plane	237
44. Systems of Higher Degree Equations	242
45. Detemine the Equations of Loci	244
46. Equations of Conic Sections and Their Use	250
47. Identifying and Graphing Polynomial and Rational Functions	252
48. Concepts of Logarithmic and Exponential Functions	254
49. Apply the Concept of Limits to Functions	257
50. Ability to Find Derivatives of Functions	260
51. Apply Derivatives to Find the Slopes of Curves, Tangents, and Normal Lines to a Curve	270
52. Identify Increasing and Decreasing Functions, Relative and Absolute Maximum and Minimum Points, Concavity, and Points of Inflection	274
53. Ability to Find Antiderivatives	282
54. Antiderivative Problems Related to Motions of Bodies	286
55. Techniques of Integration and Their Use	288
56. Integral Calculus to Find the Area Between Curves and the Volume of a Solid of Revolution	292
Sample test	315

This page intentionally left blank.

TEACHER CERTIFICATION EXAM

COMPETENCY 1.0   **COMPREHENSION OF PROPERTIES AND OPERATIONS OF THE SYSTEM OF REAL NUMBERS.**

SKILL 1.1   **Change a number in fraction form to a decimal or change a terminating or repeating decimal to a fraction.**

To convert a fraction to a decimal, simply divide the numerator (top) by the denominator (bottom). Use long division if necessary.

If a decimal has a fixed number of digits, the decimal is said to be terminating. To write such a decimal as a fraction, first determine what place value the farthest right digit is in, for example: tenths, hundredths, thousandths, ten thousandths, hundred thousands, etc. Then drop the decimal and place the string of digits over the number given by the place value.

If a decimal continues forever by repeating a string of digits, the decimal is said to be repeating. To write a repeating decimal as a fraction, follow these steps.

    a.   Let $x =$ the repeating decimal
             (ex. $x = .716716716...$)
    b.   Multiply x by the multiple of ten that will move the decimal just to the right of the repeating block of digits.
             (ex. $1000x = 716.716716...$)
    c.   Subtract the first equation from the second.
             (ex. $1000x - x = 716.716.716... - .716716...$)
    d.   Simplify and solve this equation. The repeating block of digits will subtract out.
             (ex. $999x = 716$ so $x = 716/999$)
    e.   The solution will be the fraction for the repeating decimal.

MATHEMATICS HIGH SCHOOL

SKILL 1.2 Apply the properties of real numbers: closure, commutative,
1.3 associative, distributive, properties of zero and one, and the inverse elements.

The real number properties are best explained in terms of a small set of numbers. For each property, a given set will be provided.

Axioms of Addition

Closure—For all real numbers a and b, $a + b$ is a unique real number.

Associative—For all real numbers a, b, and c, $(a + b) + c = a + (b + c)$.

Additive Identity—There exists a unique real number 0 (zero) such that $a + 0 = 0 + a = a$ for every real number a.

Additive Inverses—For each real number a, there exists a real number $-a$ (the opposite of a) such that $a + (-a) = (-a) + a = 0$.

Commutative—For all real numbers a and b, $a + b = b + a$.

Axioms of Multiplication

Closure—For all real numbers a and b, ab is a unique real number.

Associative—For all real numbers a, b, and c, $(ab)c = a(bc)$.

Multiplicative Identity—There exists a unique nonzero real number 1 (one) such that $1 \cdot a = a \cdot 1 = a$.

Multiplicative Inverses—For each nonzero rel number, there exists a real number $1/a$ (the reciprocal of a) such that $a(1/a) = (1/a)a = 1$.

Commutative—For all real numbers a and b, $ab = ba$.

The Distributive Axiom of Multiplication over Addition

For all real numbers a, b, and c, $a(b + c) = ab + ac$.

TEACHER CERTIFICATION EXAM

SKILL 1.4 **Define the real number system and its subsets.**

a. **Natural numbers**--the counting numbers, 1,2,3,...

b. **Whole numbers**--the counting numbers along with zero, 0,1,2...

c. **Integers**--the counting numbers, their opposites, and zero, ..., ⁻1,0,1,...

d. **Rationals**--all of the fractions that can be formed from the whole numbers. Zero cannot be the denominator. In decimal form, these numbers will either be terminating or repeating decimals. Simplify square roots to determine if the number can be written as a fraction.

e. **Irrationals**--real numbers that cannot be written as a fraction. The decimal forms of these numbers are neither terminating nor repeating. Examples: $\pi, e, \sqrt{2}$, etc.

f. **Real numbers**--the set of numbers obtained by combining the rationals and irrationals. Complex numbers, i.e. numbers that involve i or $\sqrt{-1}$, are not real numbers.

SKILL 1.5 **Recognize the property of denseness.**

The **Denseness Property** of real numbers states that, if all real numbers are ordered from least to greatest on a number line, there is an infinite set of real numbers between any two given numbers on the line.

Example:

Between 7.6 and 7.7, there is the rational number 7.65 in the set of real numbers.
Between 3 and 4 there exists no other natural number.

MATHEMATICS HIGH SCHOOL

SKILL 1.6 Apply the order of operations.

For standardization purposes, there is an accepted order in which operations are performed in any given algebraic expression. The following pneumonic is often used for the order in which operations are performed.

Please	Parentheses	
Excuse	Exponents	
My	Multiply	Multiply or Divide depending on which
Dear	Divide	operation is encountered first from left to right.
Aunt	Add	Add or Subtract depending on which
Sally	Subtract	operation is encountered first from left to right.

SKILL 1.7 Apply inverse operations.

Subtraction is the inverse of Addition, and vice-versa.
Division is the inverse of Multiplication, and vice-versa.
Taking a square root is the inverse of squaring, and vice-versa.

These inverse operations are used when solving equations.

SKILL 1.8 Use estimation and approximation to check reasonableness of answers.

Estimation and approximation may be used to check the reasonableness of answers.

Example: Estimate the answer.

$$\frac{58 \times 810}{1989}$$

58 becomes 60, 810 becomes 800 and 1989 becomes 2000.

$$\frac{60 \times 800}{2000} = 24$$

Word problems: An estimate may sometimes be all that is needed to solve a problem.

Example: Janet goes into a store to purchase a CD on sale for $13.95. While shopping, she sees two pairs of shoes, prices $19.95 and $14.50. She only has $50. Can she puchase everything?

Solve by rounding:

$19.95→$20.00
$14.50→$15.00
$13.95→$14.00
$49.00   Yes, she can purchase the CD and the shoes.

TEACHER CERTIFICATION EXAM

COMPETENCY 2.0   **KNOWLEDGE OF THE BASIC CONCEPTS OF SET THEORY.**

SKILL 2.1   **Apply the basic concepts of set theory.**

Set A - - $\{^-5, ^-3, 0, 1, 2, 3, 5\}$

Set B - - $\{^-7, ^-2, 0, 1, 3, 4, 5, 6\}$

Set C - - $\{^-6, ^-4, ^-2, 4, 6\}$

The **Union** ($\cup$) of two sets is the set of all the numbers which are either in the first set or the second set or both sets.

$$A \cup B \text{ is } \{^-7, ^-5, ^-3, ^-2, 0, 1, 2, 3, 4, 5, 6\}.$$

The **Intersection** ($\cap$) of two sets is the set of numbers that are in both sets.

$$A \cap B \text{ is } \{0, 1, 3, 5\}.$$

The **Null Set** is also called the empty set; it is the set that does not contain any numbers. The Null Set can be expressed two different ways: either { } or $\varnothing$. {0} is NOT the Null Set since it does have one element.

$A \cap C$ is the empty set, { }, since they do not share any elements.

TEACHER CERTIFICATION EXAM

COMPETENCY 3.0   **UNDERSTANDING OF RATIO AND PROPORTION, PERCENT, FRACTIONS, AND DECIMALS.**

SKILL 3.1   Solve real-world problems involving comparison shopping.

The unit rate for purchasing an item is its price divided by the number of pounds/ ounces, etc. in the item. The item with the lower unit rate is the lower price.

<u>Example:</u> Find the item with the best unit price:

$1.79 for 10 ounces
$1.89 for 12 ounces
$5.49 for 32 ounces

$$\frac{1.79}{10} = .179 \text{ per ounce} \quad \frac{1.89}{12} = .1575 \text{ per ounce} \quad \frac{5.49}{32} = .172 \text{ per ounce}$$

$1.89 for 12 ounces is the best price.

A second way to find the better buy is to make a proportion with the price over the number of ounces, etc. Cross multiply the proportion, writing the products above the numerator that is used. The better price will have the smaller product.

<u>Example:</u> Find the better buy:

$8.19 for 40 pounds or $4.89 for 22 pounds

Find the unit price.

$$\frac{40}{8.19} = \frac{1}{x} \qquad\qquad \frac{22}{4.89} = \frac{1}{x}$$
$$40x = 8.19 \qquad\qquad 22x = 4.89$$
$$x = .20475 \qquad\qquad x = .222\overline{27}$$

Since $.20475 < .222\overline{27}$, $8.19 is less and is a better buy.

MATHEMATICS HIGH SCHOOL

TEACHER CERTIFICATION EXAM

SKILL 3.2 Solve real-world problems involving purchases and a rate of sales tax.

To find the amount of sales tax on an item, change the percent of sales tax into an equivalent decimal number. Then multiply the decimal number times the price of the object to find the sales tax. The total cost of an item will be the price of the item plus the sales tax.

Example: A guitar costs $120 plus 7% sales tax. How much are the sales tax and the total bill?

$$7\% = .07 \text{ as a decimal } (.07)(120) = \$8.40 \text{ sales tax}$$
$$\$120 + \$8.40 = \$128.40 \leftarrow \text{total price}$$

Example: A suit costs $450 plus 6½% sales tax. How much are the sales tax and the total bill?

$$6\tfrac{1}{2}\% = .065 \text{ as a decimal}$$
$$(.065)(450) = \$29.25 \text{ sales tax}$$
$$\$450 + \$29.25 = \$479.25 \leftarrow \text{total price}$$

SKILL 3.3 Solve word problems involving operations of ratio and proportions, percents, decimals, or fractions.

A **ratio** is a comparison of 2 numbers. If a class had 11 boys and 14 girls, the ratio of boys to girls could be written one of 3 ways:

$$11:14 \quad \text{or} \quad 11 \text{ to } 14 \quad \text{or} \quad \frac{11}{14}$$

The ratio of girls to boys is:

$$14:11, \; 14 \text{ to } 11 \; \text{or} \; \frac{14}{11}$$

Ratios can be reduced when possible. A ratio of 12 cats to 18 dogs would reduce to 2:3, 2 to 3 or $2/3$.

Note: Read ratio questions carefully. Given a group of 6 adults and 5 children, the ratio of children to the entire group would be 5:11.

A **proportion** is an equation in which a fraction is set equal to another. To solve the proportion, multiply each numerator times the other fraction's denominator. Set these two products equal to each other and solve the resulting equation. This is called **cross-multiplying** the proportion.

Example: $\dfrac{4}{15} = \dfrac{x}{60}$ is a proportion.

To solve this, cross multiply.

$$(4)(60) = (15)(x)$$
$$240 = 15x$$
$$16 = x$$

Example: $\dfrac{x+3}{3x+4} = \dfrac{2}{5}$ is a proportion.

To solve, cross multiply.

$$5(x+3) = 2(3x+4)$$
$$5x + 15 = 6x + 8$$
$$7 = x$$

Example: $\dfrac{x+2}{8} = \dfrac{2}{x-4}$ is another proportion.

To solve, cross multiply.

$$(x+2)(x-4) = 8(2)$$
$$x^2 - 2x - 8 = 16$$
$$x^2 - 2x - 24 = 0$$
$$(x-6)(x+4) = 0$$
$$x = 6 \text{ or } x = {}^-4$$

Both answers work.

Fractions, decimals, and percents can be used interchangeably within problems.

→ To change a percent into a decimal, move the decimal point two places to the left and drop off the percent sign.

→ To change a decimal into a percent, move the decimal two places to the right and add on a percent sign.

→ To change a fraction into a decimal, divide the numerator by the denominator.

→ To change a decimal number into an equivalent fraction, write the decimal part of the number as the fraction's numerator. As the fraction's denominator use the place value of the last column of the decimal. Reduce the resulting fraction as far as possible.

Example: J.C. Nickels has Hunch jeans 1/4 off the usual price of $36.00. Shears and Roadkill have the same jeans 30% off their regular price of $40. Find the cheaper price.

1/4 = .25 so .25(36) = $9.00 off $36 - 9 = $27 sale price

30% = .30 so .30(40) = $12 off $40 - 12 = $28 sale price

The price at Shears and Roadkill is actually lower.

TEACHER CERTIFICATION EXAM

SKILL 3.4 Solve problems involving elapsed time.

Elapsed time problems are usually one of two types. One type of problem is the elapsed time between 2 times given in hours, minutes, and seconds. The other common type of problem is between 2 times given in months and years.

For any time of day past noon, change it into military time by adding 12 hours. For instance, 1:15 p.m. would be 13:15. Remember when you borrow a minute or an hour in a subtraction problem that you have borrowed 60 more seconds or minutes.

Example: Find the time from 11:34:22 a.m. until 3:28:40 p.m.

    First change 3:28:40 p.m. to 15:28:40 p.m.
    Now subtract   - 11:34:22 a.m.
                                              :18

Borrow an hour and add 60 more minutes. Subtract
    14:88:40 p.m.
    - 11:34:22 a.m.
      3:54:18 ↔ 3 hours, 54 minutes, 18 seconds

Example: John lived in Arizona from September 91 until March 95. How long is that?

		year	month
March 95	=	95	03
September 91	= -	91	09

Borrow a year, change it into 12 more months, and subtract.

		year	month
March 95	=	94	15
September 91	= -	91	09
		3 yr	6 months

Example: A race took the winner 1 hr. 58 min. 12 sec. on the first half of the race and 2 hr. 9 min. 57 sec. on the second half of the race. How much time did the entire race take?

      1 hr. 58 min. 12 sec.
    + 2 hr. 9 min. 57 sec.   Add these numbers
      3 hr. 67 min. 69 sec.
    + 1 min -60 sec.   Change 60 seconds to 1 min.
      3 hr. 68 min. 9 sec.
    + 1 hr.-60 min.   .   Change 60 minutes to 1 hr.
      4 hr. 8 min. 9 sec. ←final answer

MATHEMATICS HIGH SCHOOL

TEACHER CERTIFICATION EXAM

SKILL 3.5 **Order of Operations.**

The Order of Operations are to be followed when evaluating algebraic expressions. Follow these steps in order:

1. Simplify inside grouping characters such as parentheses, brackets, square root, fraction bar, etc.

2. Multiply out expressions with exponents.

3. Do multiplication or division, from left to right.

4. Do addition or subtraction, from left to right.

Samples of simplifying expressions with exponents:

$$(^-2)^3 = -8 \qquad ^-2^3 = \,^-8$$
$$(^-2)^4 = 16 \qquad ^-2^4 = 16 \qquad \text{Note change of sign.}$$
$$(2/3)^3 = 8/27$$
$$5^0 = 1$$
$$4^{-1} = 1/4$$

MATHEMATICS HIGH SCHOOL

TEACHER CERTIFICATION EXAM

COMPETENCY 4.0   **COMPREHENSION OF NUMBER THEORY.**

SKILL 4.1   **Apply prime factorization**

Prime numbers are numbers that can only be factored into 1 and the number itself. When factoring into prime factors, all the factors must be numbers that cannot be factored again (without using 1). Initially numbers can be factored into any 2 factors. Check each resulting factor to see if it can be factored again. Continue factoring until all remaining factors are prime. This is the list of prime factors. Regardless of what way the original number was factored, the final list of prime factors will always be the same.

Example: Factor 30 into prime factors.

    Factor 30 into any 2 factors.
       5 · 6   Now factor the 6.
      5 · 2 · 3   These are all prime factors.

    Factor 30 into any 2 factors.
       3 · 10   Now factor the 10.
      3 · 2 · 5   These are the same prime factors even though the original factors were different.

Example: Factor 240 into prime factors.

    Factor 240 into any 2 factors.
      24 · 10   Now factor both 24 and 10.
      4 · 6 · 2 · 5   Now factor both 4 and 6.
    2 · 2 · 2 · 3 · 2 · 5   These are prime factors.

This can also be written as $2^4 \cdot 3 \cdot 5$.

SKILL 4.2 Define GCF and LCM.

GCF is the abbreviation for the **greatest common factor**. The GCF is the largest number that is a factor of all the numbers given in a problem. The GCF can be no larger than the smallest number given in the problem. If no other number is a common factor, then the GCF will be the number 1. To find the GCF, list all possible factors of the smallest number given (include the number itself). Starting with the largest factor (which is the number itself), determine if it is also a factor of all the other given numbers. If so, that is the GCF. If that factor doesn't work, try the same method on the next smaller factor. Continue until a common factor is found. That is the GCF. Note: There can be other common factors besides the GCF.

Example: Find the GCF of 12, 20, and 36.

The smallest number in the problem is 12. The factors of 12 are 1,2,3,4,6 and 12. 12 is the largest factor, but it does not divide evenly into 20. Neither does 6, but 4 will divide into both 20 and 36 evenly. Therefore, 4 is the GCF.

Example: Find the GCF of 14 and 15.

Factors of 14 are 1,2,7 and 14. 14 is the largest factor, but it does not divide evenly into 15. Neither does 7 or 2. Therefore, the only factor common to both 14 and 15 is the number 1, the GCF.

LCM is the abbreviation for **least common multiple**. The least common multiple of a group of numbers is the smallest number that all of the given numbers will divide into. The least common multiple will always be the largest of the given numbers or a multiple of the largest number.

Example: Find the LCM of 20, 30 and 40.

The largest number given is 40, but 30 will not divide evenly into 40. The next multiple of 40 is 80 (2 x 40), but 30 will not divide evenly into 80 either. The next multiple of 40 is 120. 120 is divisible by both 20 and 30, so 120 is the LCM (least common multiple).

Example: Find the LCM of 96, 16 and 24.

The largest number is 96. 96 is divisible by both 16 and 24, so 96 is the LCM.

TEACHER CERTIFICATION EXAM

SKILL 4.3 Explain divisibility tests and why they work (divisors 2,3,4,5,6,8,9,10).

a. A number is divisible by 2 if that number is an even number (which means it ends in 0,2,4,6 or 8).

1,354 ends in 4, so it is divisible by 2. 240,685 ends in a 5, so it is not divisible by 2.

b. A number is divisible by 3 if the sum of its digits is evenly divisible by 3.

The sum of the digits of 964 is 9+6+4 = 19. Since 19 is not divisible by 3, neither is 964. The digits of 86,514 is 8+6+5+1+4 = 24. Since 24 is divisible by 3, 86,514 is also divisible by 3.

c. A number is divisible by 4 if the number in its last 2 digits is evenly divisible by 4.

The number 113,336 ends with the number 36 in the last 2 columns. Since 36 is divisible by 4, then 113,336 is also divisible by 4.

The number 135,627 ends with the number 27 in the last 2 columns. Since 27 is not evenly divisible by 4, then 135,627 is also not divisible by 4.

d. A number is divisible by 5 if the number ends in either a 5 or a 0.

225 ends with a 5 so it is divisible by 5. The number 470 is also divisible by 5 because its last digit is a 0. 2,358 is not divisible by 5 because its last digit is an 8, not a 5 or a 0.

e. A number is divisible by 6 if the number is even and the sum of its digits is evenly divisible by 3.

4,950 is an even number and its digits add to 18. (4+9+5+0 = 18) Since the number is even and the sum of its digits is 18 (which is divisible by 3), then 4950 is divisible by 6. 326 is an even number, but its digits add up to 11. Since 11 is not divisible by 3, then 326 is not divisible by 6. 698,135 is not an even number, so it cannot possibly be divided evenly by 6.

MATHEMATICS HIGH SCHOOL

f. A number is divisible by 8 if the number in its last 3 digits is evenly divisible by 8.

The number 113,336 ends with the 3-digit number 336 in the last 3 places. Since 336 is divisible by 8, then 113,336 is also divisible by 8.
The number 465,627 ends with the number 627 in the last 3 places. Since 627 is not evenly divisible by 8, then 465,627 is also not divisible by 8.

g. A number is divisible by 9 if the sum of its digits is evenly divisible by 9.

The sum of the digits of 874 is 8+7+4 = 19. Since 19 is not divisible by 9, neither is 874. The digits of 116,514 is 1+1+6+5+1+4 = 18. Since 18 is divisible by 9, 116,514 is also divisible by 9.

h. A number is divisible by 10 if the number ends in the digit 0.

305 ends with a 5 so it is not divisible by 10. The number 2,030,270 is divisible by 10 because its last digit is a 0. 42,978 is not divisible by 10 because its last digit is an 8, not a 0.

i. Why these rules work.

All even numbers are divisible by 2 by definition. A 2-digit number (with T as the tens digit and U as the ones digit) has as its sum of the digits, T + U. Suppose this sum of T + U is divisible by 3. Then it equals 3 times some constant, K. So, T + U = 3K. Solving this for U, U = 3K - T. The original 2 digit number would be represented by 10T + U. Substituting 3K - T in place of U, this 2-digit number becomes
10T + U = 10T + (3K - T) = 9T + 3K. This 2-digit number is clearly divisible by 3, since each term is divisible by 3. Therefore, if the sum of the digits of a number is divisible by 3, then the number itself is also divisible by 3. Since 4 divides evenly into 100, 200, or 300, 4 will divide evenly into any amount of hundreds. The only part of a number that determines if 4 will divide into it evenly is the number in the last 2 places. Numbers divisible by 5 end in 5 or 0. This is clear if you look at the answers to the multiplication table for 5. Answers to the multiplication table for 6 are all even numbers. Since 6 factors into 2 times 3, the divisibility rules for 2 and 3 must both work. Any number of thousands is divisible by 8. Only the last 3 places of the number determine whether or not it is divisible by 8. A 2 digit number (with T as the tens digit and U as the ones digit) has as its sum of the digits, T + U. Suppose this sum of T + U is divisible by 9. Then it equals 9 times some constant, K. So, T + U = 9K. Solving this for U, U = 9K - T. The original 2-digit number would be represented by 10T + U. Substituting 9K - T in place of U, this 2-digit number becomes
10T + U = 10T + (9K - T) = 9T + 9K. This 2-digit number is clearly divisible by 9, since each term is divisible by 9. Therefore, if the sum of the digits

of a number is divisible by 9, then the number itself is also divisible by 9. Numbers divisible by 10 must be multiples of 10 which all end in a zero.

SKILL 4.4 **Define prime and composite numbers.**

Prime numbers are whole numbers greater than 1 that have only 2 factors, 1 and the number itself. Examples of prime numbers are 2,3,5,7,11,13,17, or 19. Note that 2 is the only even prime number.

Composite numbers are whole numbers that have more than 2 different factors. For example 9 is composite because besides factors of 1 and 9, 3 is also a factor. 70 is also composite because besides the factors of 1 and 70, the numbers 2,5,7,10,14, and 35 are also all factors.

Remember that the number 1 is neither prime nor composite.

SKILL 4.5 **Use the laws of exponents to evaluate expressions.**

The **exponent form** is a shortcut method to write repeated multiplication. The **base** is the factor. The **exponent** tells how many times that number is multiplied by itself.

The following are basic rules for exponents:

- $a^1 = a$ for all values of a; thus $17^1 = 17$
- $b^0 = 1$ for all values of b; thus $24^0 = 1$
- $10^n = 1$ with n zeros; thus $10^6 = 1{,}000{,}000$

TEACHER CERTIFICATION EXAM

SKILL 4.6 Use scientific notation.

To change a number into scientific notation, move the decimal point so that only one number from 1 to 9 is in front of the decimal point. Drop off any trailing zeros. Multiply this number times 10 to a power. The power is the number of positions that the decimal point is moved. The power is negative if the original number is a decimal number between 1 and -1. Otherwise the power is positive.

Example: Change into scientific notation:

4,380,000,000	Move decimal behind the 4
4.38	Drop trailing zeros.
$4.38 \times 10^?$	Count positions that the decimal point has moved.
4.38×10^9	This is the answer.
$^-.0000407$	Move decimal behind the 4
$^-4.07$	Count positions that the decimal point has moved.
$^-4.07 \times 10^{-5}$	Note negative exponent.

If a number is already in scientific notation, it can be changed back into the regular decimal form. If the exponent on the number 10 is negative, move the decimal point to the left. If the exponent on the number 10 is positive, move the decimal point to the right that number of places.

Example: Change back into decimal form:

3.448×10^{-2}	Move decimal point 2 places left, since exponent is negative.
.03448	This is the answer.
6×10^4	Move decimal point 4 places right, since exponent is negative.
60,000	This is the answer.

To add or subtract in scientific notation, the exponents must be the same. Then add the decimal portions, keeping the power of 10 the same. Then move the decimal point and adjust the exponent to keep the number in front of the decimal point from 1 to 9.

MATHEMATICS HIGH SCHOOL

Example:

6.22×10^3
$+ 7.48 \times 10^3$ Add these as is.
─────────────
13.70×10^3 Now move decimal 1 more place to the left and
1.37×10^4 add 1 more exponent.

To multiply or divide in scientific notation, multiply or divide the decimal part of the numbers. In multiplication, add the exponents of 10. In division, subtract the exponents of 10. Then move the decimal point and adjust the exponent to keep the number in front of the decimal point from 1 to 9.

Example:

$(5.2 \times 10^5)(3.5 \times 10^2)$ Multiply $5.2 \cdot 3.5$

18.2×10^7 Add exponent

1.82×10^8 Move decimal point and increase the exponent by 1.

Example:

$\dfrac{(4.1076 \times 10^3)}{2.8 \times 10^{-4}}$ Divide 4.1076 by 2.8

Subtract $3 - (^-4)$

1.467×10^7

TEACHER CERTIFICATION EXAM

COMPETENCY 5.0   KNOWLEDGE OF GEOMETRIC CONCEPTS, FORMULAS, AND CONSTRUCTIONS, INCLUDING RIGHT TRIANGLE RELATIONSHIPS.

SKILL 5.1   Compute the area remaining when sections are cut out of a given figure composed of triangles, squares, rectangles, parallelograms, trapezoids, or circles.

The strategy for solving problems of this nature should be to identify the given shapes and choose the correct formulas. Subtract the smaller cut out shape from the larger shape.

Sample problems:

1. Find the area of one side of the metal in the circular flat washer shown below:

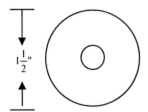

1. the shapes are both circles.

2. use the formula $A = \pi r^2$ for both.

(Inside diameter is 3/8")

Area of larger circle	Area of smaller circle
$A = \pi r^2$	$A = \pi r^2$
$A = \pi(.75^2)$	$A = \pi(.1875^2)$
$A = 1.76625$ in^2	$A = .1104466$ in^2

Area of metal washer = larger area - smaller area

$$= 1.76625 \text{ in}^2 - .1104466 \text{ in}^2$$
$$= 1.6558034 \text{ in}^2$$

TEACHER CERTIFICATION EXAM

2. You have decided to fertilize your lawn. The shapes and dimensions of your lot, house, pool and garden are given in the diagram below. The shaded area will not be fertilized. If each bag of fertilizer costs $7.95 and covers 4,500 square feet, find the total number of bags needed and the total cost of the fertilizer.

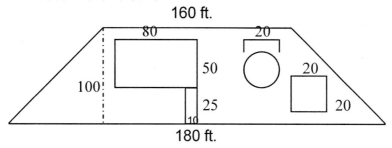

Area of Lot
$A = \frac{1}{2} h(b_1 + b_2)$
$A = \frac{1}{2}(100)(180 + 160)$
$A = 17,000$ sq ft

Area of House
$A = LW$
$A = (80)(50)$
$A = 4,000$ sq ft

Area of Driveway
$A = LW$
$A = (10)(25)$
$A = 250$ sq ft

Area of Pool
$A = \pi r^2$
$A = \pi(10)^2$
$A = 314.159$ sq. ft.

Area of Garden
$A = s^2$
$A = (20)^2$
$A = 400$ sq. ft.

Total area to fertilize = Lot area - (House + Driveway + Pool + Garden)
 = 17,000 - (4,000 + 250 + 314.159 + 400)
 = 12,035.841 sq ft

Number of bags needed = Total area to fertilize / 4,500 sq.ft. bag
 = 12,035.841 / 4,500
 = 2.67 bags

Since we cannot purchase 2.67 bags we must purchase 3 full bags.

Total cost = Number of bags * $7.95
 = 3 * $7.95
 = $23.85

SKILL 5.2 Determine the change in the area or volume of a plane figure when its dimensions are altered.

Examining the change in area or volume of a given figure requires first to find the existing area given the original dimensions and then finding the new area given the increased dimensions.

Sample problem:

Given the rectangle below determine the change in area if the length is increase by 5 and the width is increased by 7.

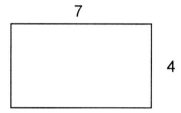

Draw and label a sketch of the new rectangle.

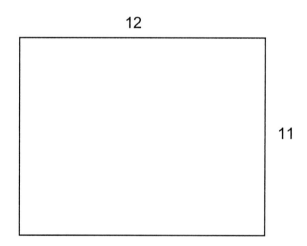

Find the areas.

Area of original = LW   Area of enlarged shape = LW
         = (7)(4)   = (12)(11)
         = 28 units^2 = 132 units^2

The change in area is 132 − 28 = 104 units^2.

SKILL 5.3 Find the area of compound shapes.

Cut the compound shape into smaller, more familiar shapes and then compute the total area by adding the areas of the smaller parts.

Sample problem:

Find the area of the given shape.

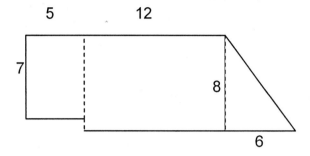

1. Using a dotted line we have cut the shape into smaller parts that are familiar.

2. Use the appropriate formula for each shape and find the sum of all areas.

Area 1 = LW   Area 2 = LW   Area 3 = ½bh
   = (5)(7)   = (12)(8)   = ½(6)(8)
   = 35 units^2 = 96 units^2 = 24 units^2

Total area = Area 1 + Area 2 + Area 3
         = 35 + 96 + 24
         = 155 units^2

SKILL 5.4 Estimate measurements of familiar objects.

It is necessary to be familiar with the metric and customary system in order to estimate measurements.

Some common equivalents include:

ITEM	APPROXIMATELY EQUAL TO	
	METRIC	IMPERIAL
large paper clip	1 gram	1 ounce
1 quart	1 liter	
average sized man	75 kilograms	170 pounds
1 yard	1 meter	
math textbook	1 kilogram	2 pounds
1 mile	1 kilometer	
1 foot	30 centimeters	
thickness of a dime	1 millimeter	0.1 inches

Estimate the measurement of the following items:

The length of an adult cow = _____ meters
The thickness of a compact disc = _____ millimeters
Your height = _____ meters
length of your nose = _____ centimeters
weight of your math textbook = _____ kilograms
weight of an automobile = _____ kilograms
weight of an aspirin = _____ grams

SKILL 5.5 Measure objects to the nearest given unit.

Given a set of objects and their measurements, the use of rounding procedures is helpful when attempting to round to the nearest given unit. When rounding to a given place value, it is necessary to look at the number in the next smaller place. If this number is 5 or more, the number in the place we are rounding to is increased by one and all numbers to the right are changed to zero. If the number is less than 5, the number in the place we are rounding to stays the same and all numbers to the right are changed to zero.

One method of rounding measurements can require an additional step. First, the measurement must be converted to a decimal number. Then the rules for rounding applied.

Sample problem:

1. Round the measurements to the given units.

MEASUREMENT	ROUND TO NEAREST	ANSWER
1 foot 7 inches	foot	2 ft
5 pound 6 ounces	pound	5 pounds
5 9/16 inches	inch	6 inches

Solution:

Convert each measurement to a decimal number. Then apply the rules for rounding.

1 foot 7 inches = $1\frac{7}{12}$ ft = 1.58333 ft, round up to 2 ft

5 pounds 6 ounces = $5\frac{6}{16}$ pounds = 5.375 pound, round to 5 pounds

$5\frac{9}{16}$ inches = 5.5625 inches, round up to 6 inches

TEACHER CERTIFICATION EXAM

SKILL 5.6 Make conversions within a system.

There are many methods for converting measurements within a system. One method is to multiply the given measurement by a conversion factor. This conversion factor is the ratio of:

$$\frac{\text{new units}}{\text{old units}} \quad \text{OR} \quad \frac{\text{what you want}}{\text{what you have}}$$

Sample problems:

1. Convert 3 miles to yards.

$$\frac{3 \text{ miles}}{1} \times \frac{1{,}760 \text{ yards}}{1 \text{ mile}} = \frac{\text{yards}}{}$$

$$= 5{,}280 \text{ yards}$$

1. multiply by the conversion factor
2. cancel the miles units
3. solve

2. Convert 8,750 meters to kilometers.

$$\frac{8{,}750 \text{ meters}}{1} \times \frac{1 \text{ kilometer}}{1000 \text{ meters}} = \frac{\text{km}}{}$$

$$= 8.75 \text{ kilometers}$$

1. multiply by the conversion factor
2. cancel the meters units
3. solve

MATHEMATICS HIGH SCHOOL

SKILL 5.7 Explain greatest possible error, accuracy, and precision.

Most numbers in mathematics are "exact" or "counted". Measurements are "approximate". They usually involve interpolation or figuring out which mark on the ruler is closest. Any measurement you get with a measuring device is approximate. Variations in measurement are called precision and accuracy.

Precision is a measurement of how exactly a measurement is made, without reference to a true or real value. If a measurement is precise it can be made again and again with little variation in the result. The precision of a measuring device is the smallest fractional or decimal division on the instrument. The smaller the unit or fraction of a unit on the measuring device, the more precisely it can measure.

The greatest possible error of measurement is always equal to one-half the smallest fraction of a unit on the measuring device.

Accuracy is a measure of how close the result of measurement comes to the "true" value.

If you are throwing darts, the true value is the bull's eye. If the three darts land on the bull's eye, the dart thrower is both precise (all land near the same spot) and accurate (the darts all land on the "true" value).
The greatest measure of error allowed is called the tolerance. The least acceptable limit is called the lower limit and the greatest acceptable limit is called the upper limit. The difference between the upper and lower limits is called the tolerance interval. For example, a specification for an automobile part might be 14.625 ± 0.005 mm. This means that the smallest acceptable length of the part is 14.620 mm and the largest length acceptable is 14.630 mm. The tolerance interval is 0.010 mm. One can see how it would be important for automobile parts to be within a set of limits in terms of length. If the part is too long or too short it will not fit properly and vibrations will occur weakening the part and eventually causing damage to other parts.

TEACHER CERTIFICATION EXAM

COMPETENCY 6.0   KNOWLEDGE OF BASIC CONCEPTS OF PROBABILITY AND STATISTICS AS APPLIED TO EDUCATIONAL DATA.

SKILL 6.1   Interpret graphical data involving measures of location (percentiles, stanines and quartiles).

Percentiles divide data into 100 equal parts. A person whose score falls in the 65th percentile has outperformed 65 percent of all those who took the test. This does not mean that the score was 65 percent out of 100 nor does it mean that 65 percent of the questions answered were correct. It means that the grade was higher than 65 percent of all those who took the test.

Stanine "standard nine" scores combine the understandability of percentages with the properties of the normal curve of probability. Stanines divide the bell curve into nine sections, the largest of which stretches from the 40th to the 60th percentile and is the "Fifth Stanine" (the average of taking into account error possibilities).

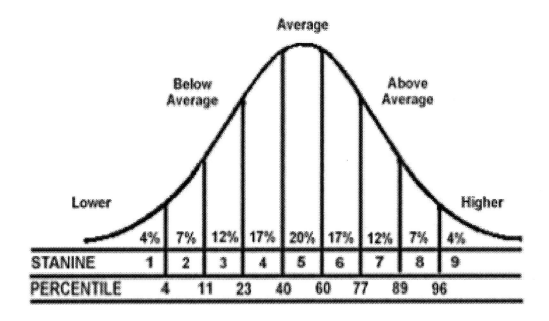

Quartiles divide the data into 4 parts. First find the median of the data set (Q2), then find the median of the upper (Q3) and lower (Q1) halves of the data set. If there are an odd number of values in the data set, include the median value in both halves when finding quartile values. For example, given the data set: {1, 4, 9, 16, 25, 36, 49, 64, 81} first find the median value, which is 25 this is the second quartile. Since there are an odd number of values in the data set (9), we include the median in both halves. To find the quartile values, we much find the medians of: {1, 4, 9, 16, 25}

and {25, 36, 49, 64, 81}. Since each of these subsets had an odd number of elements (5), we use the middle value. Thus the first quartile value is 9 and the third quartile value is 49. If the data set had an even number of elements, average the middle two values. The quartile values are always either one of the data points, or exactly half way between two data points.

Sample problem:

1. Given the following set of data, find the percentile of the score 104.
    70, 72, 82, 83, 84, 87, 100, 104, 108, 109, 110, 115

Solution: Find the percentage of scores below 104.

7/12 of the scores are less than 104. This is 58.333%; therefore, the score of 104 is in the 58th percentile.

2. Find the first, second and third quartile for the data listed.
    6, 7, 8, 9, 10, 12, 13, 14, 15, 16, 18, 23, 24, 25, 27, 29, 30, 33, 34, 37

Quartile 1: The 1st Quartile is the median of the lower half of the data set, which is 11.

Quartile 2: The median of the data set is the 2nd Quartile, which is 17.

Quartile 3: The 3rd Quartile is the median of the upper half of the data set, which is 28.

SKILL 6.2 Determine final probabilities of dependent or independent events.

Dependent events occur when the probability of the second event depends on the outcome of the first event. For example, consider the two events (A) it is sunny on Saturday and (B) you go to the beach. If you intend to go to the beach on Saturday, rain or shine, then A and B may be independent. If however, you plan to go to the beach only if it is sunny, then A and B may be dependent. In this situation, the probability of event B will change depending on the outcome of event A.

Suppose you have a pair of dice, one red and one green. If you roll a three on the red die and then roll a four on the green die, we can see that these events do not depend on the other. The total probability of the two independent events can be found by multiplying the separate probabilities.

$$P(A \text{ and } B) = P(A) \times P(B)$$

$$= 1/6 \times 1/6$$
$$= 1/36$$

Many times, however, events are not independent. Suppose a jar contains 12 red marbles and 8 blue marbles. If you randomly pick a red marble, replace it and then randomly pick again, the probability of picking a red marble the second time remains the same. However, if you pick a red marble, and then pick again without replacing the first red marble, the second pick becomes dependent upon the first pick.

P(Red and Red) with replacement = P(Red) × P(Red)
$$= 12/20 \times 12/20$$
$$= 9/25$$

P(Red and Red) without replacement = P(Red) × P(Red)
$$= 12/20 \times 11/19$$
$$= 33/95$$

SKILL 6.3 Predict odds in favor of a given outcome.

Odds are defined as the ratio of the number of favorable outcomes to the number of unfavorable outcomes. The sum of the favorable outcomes and the unfavorable outcomes should always equal the total possible outcomes.

For example, given a bag of 12 red and 7 green marbles compute the odds of randomly selecting a red marble.

$$\text{Odds of red} = \frac{12}{19} : \frac{7}{19} \text{ or } 12:7.$$

$$\text{Odds of not getting red} = \frac{7}{19} : \frac{12}{19} \text{ or } 7:12.$$

In the case of flipping a coin, it is equally likely that a head or a tail will be tossed. The odds of tossing a head are 1:1. This is called even odds.

TEACHER CERTIFICATION EXAM

SKILL 6.4 **Compute the mean, median, mode, and range.**

Mean, median and mode are three measures of central tendency. The **mean** is the average of the data items. The **median** is found by putting the data items in order from smallest to largest and selecting the item in the middle (or the average of the two items in the middle). The **mode** is the most frequently occurring item.

Range is a measure of variability. It is found by subtracting the smallest value from the largest value.

Sample problem:

Find the mean, median, mode and range of the test score listed below:

85	77	65
92	90	54
88	85	70
75	80	69
85	88	60
72	74	95

Mean (X) = sum of all scores ÷ number of scores
= 78

Median = put numbers in order from smallest to largest. Pick middle number.
54, 60, 65, 69, 70, 72, 74, 75, 77, 80, 85, 85, 85, 88, 88, 90, 92, 95
                                          -- --
                                     both in middle
Therefore, median is average of two numbers in the middle or 78.5

Mode = most frequent number
= 85

Range = largest number minus the smallest number
= 95 − 54
= 41

MATHEMATICS HIGH SCHOOL

SKILL 6.5 Determine whether the mean, median or mode is the best measure of central tendency in a given situation.

Different situations require different information. If we examine the circumstances under which an ice cream store owner may use statistics collected in the store, we find different uses for different information.

Over a 7-day period, the store owner collected data on the ice cream flavors sold. He found the mean number of scoops sold was 174 per day. The most frequently sold flavor was vanilla. This information was useful in determining how much ice cream to order in all and in what amounts for each flavor.

In the case of the ice cream store, the median and range had little business value for the owner.

Consider the set of test scores from a math class: 0, 16, 19, 65, 65, 65, 68, 69, 70, 72, 73, 73, 75, 78, 80, 85, 88, and 92. The mean is 64.06 and the median is 71. Since there are only three scores less than the mean out of the eighteen score, the median (71) would be a more descriptive score.

Retail store owners may be most concerned with the most common dress size so they may order more of that size than any other.

SKILL 6.6 Apply basic statistical concepts (noncomputational),

Basic statistical concepts can be applied without computations. For example, inferences can be drawn from a graph or statistical data. A bar graph could display which grade level collected the most money. Student test scores would enable the teacher to determine which units need to be remediated.

TEACHER CERTIFICATION EXAM

COMPETENCY 7.0 ABILITY TO INTERPRET, CONSTRUCT, INTERPOLATE, AND EXTRAPOLATE FROM CHARTS, TABLES, AND GRAPHS.

SKILL 7.1 Interpolate information.

To make a **bar graph** or a **pictograph**, determine the scale to be used for the graph. Then determine the length of each bar on the graph or determine the number of pictures needed to represent each item of information. Be sure to include an explanation of the scale in the legend.

Example: A class had the following grades:
 4 A's, 9 B's, 8 C's, 1 D, 3 F's.
 Graph these on a bar graph and a pictograph.

Pictograph

Grade	Number of Students
A	☺☺☺☺
B	☺☺☺☺☺☺☺☺☺
C	☺☺☺☺☺☺☺☺
D	☺
F	☺☺☺

Bar graph

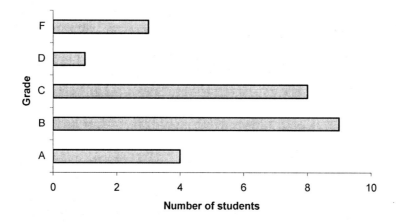

MATHEMATICS HIGH SCHOOL

To make a **line graph**, determine appropriate scales for both the vertical and horizontal axes (based on the information to be graphed). Describe what each axis represents and mark the scale periodically on each axis. Graph the individual points of the graph and connect the points on the graph from left to right.

Example: Graph the following information using a line graph.

The number of National Merit finalists/school year

	90-'91	91-'92	92-'93	93-'94	94-'95	95-'96
Central	3	5	1	4	6	8
Wilson	4	2	3	2	3	2

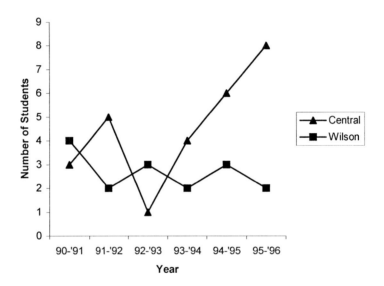

To make a **circle graph**, total all the information that is to be included on the graph. Determine the central angle to be used for each sector of the graph using the following formula:

$$\frac{\text{information}}{\text{total information}} \times 360° = \text{degrees in central} \sphericalangle$$

Lay out the central angles to these sizes, label each section and include its percent.

TEACHER CERTIFICATION EXAM

Example: Graph this information on a circle graph:

Monthly expenses:

Rent, $400
Food, $150
Utilities, $75
Clothes, $75
Church, $100
Misc., $200

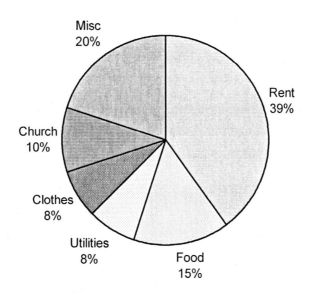

SKILL 7.2   Interpret information from bar, line, picto-, and circle graphs.

To read a bar graph or a pictograph, read the explanation of the scale that was used in the legend. Compare the length of each bar with the dimensions on the axes and calculate the value each bar represents. On a pictograph count the number of pictures used in the chart and calculate the value of all the pictures.

To read a circle graph, find the total of the amounts represented on the entire circle graph. To determine the actual amount that each sector of the graph represents, multiply the percent in a sector times the total amount number.

To read a chart read the row and column headings on the table. Use this information to evaluate the given information in the chart.

MATHEMATICS HIGH SCHOOL

TEACHER CERTIFICATION EXAM

COMPETENCY 8.0 **KNOWLEDGE OF ALGEBRAIC FUNCTIONS, RELATIONS, AND THEIR GRAPHS.**

SKILL 8.1 Determine which relations are functions, given mappings, sets of ordered pairs, rules, and graphs.

- A **relation** is any set of ordered pairs.

- The **domain** of a relation is the set made of all the first coordinates of the ordered pairs.

- The **range** of a relation is the set made of all the second coordinates of the ordered pairs.

- A **function** is a relation in which different ordered pairs have different first coordinates. (No x values are repeated.)

- A **mapping** is a diagram with arrows drawn from each element of the domain to the corresponding elements of the range. If 2 arrows are drawn from the same element of the domain, then it is not a function.

- On a graph, use the **vertical line test** to look for a function. If any vertical line intersects the graph of a relation in more than one point, then the relation is not a function.

1. Determine the domain and range of this mapping.

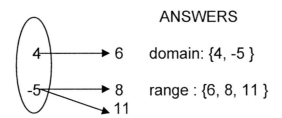

ANSWERS

domain: {4, -5 }

range : {6, 8, 11 }

MATHEMATICS HIGH SCHOOL

2. Determine which of these are functions:

   a. $\{(1,^-4),(27,1)(94,5)(2,^-4)\}$
   b. $f(x) = 2x - 3$
   c. $A = \{(x,y) \mid xy = 24\}$
   d. $y = 3$
   e. $x = {}^-9$
   f. $\{(3,2),(7,7),(0,5),(2,^-4),(8,^-6),(1,0),(5,9),(6,^-4)\}$

3. Determine the domain and range of this graph.

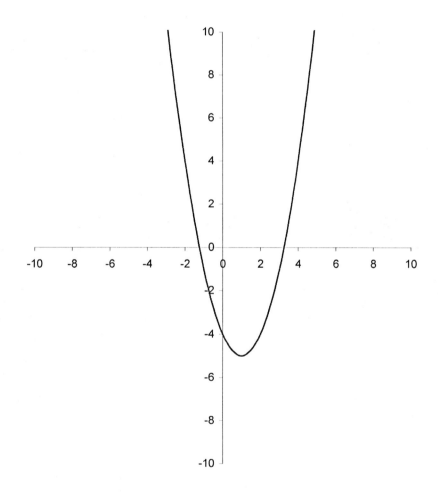

SKILL 8.2 Define a relation and state its domain and range.

- A **relation** is any set of ordered pairs.
- The **domain** of the relation is the set of all first co-ordinates of the ordered pairs. (These are the x coordinates.)
- The **range** of the relation is the set of all second co-ordinates of the ordered pairs. (These are the y coordinates.)

1. If $A = \{(x,y) \mid y = x^2 - 6\}$, find the domain and range.

2. Give the domain and range of set B if:

   $B = \{(1, ^-2),(4, ^-2),(7, ^-2),(6, ^-2)\}$

3. Determine the domain of this function:

   $$f(x) = \frac{5x+7}{x^2-4}$$

4. Determine the domain and range of these graphs.

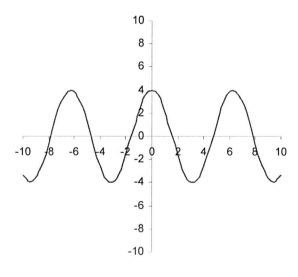

MATHEMATICS HIGH SCHOOL

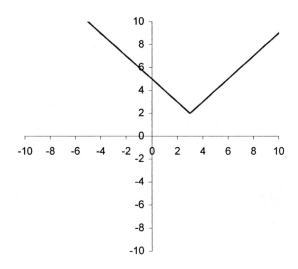

5. If $E = \{(x,y) \mid y = 5\}$, find the domain and range.

6. Determine the ordered pairs in the relation shown in this mapping.

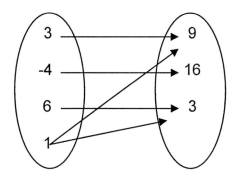

SKILL 8.3 Identify and graph special functions (absolute value, greatest integer, and identity).

-The **absolute value function** for a 1st degree equation is of the form: $y = m(x - h) + k$. Its graph is in the shape of a $\vee$. The point (h,k) is the location of the maximum/minimum point on the graph. "± m" are the slopes of the 2 sides of the $\vee$. The graph opens up if m is positive and down if m is negative.

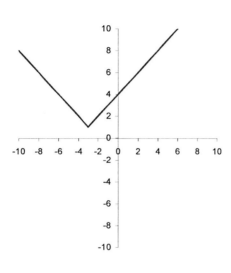

$y = |x + 3| + 1$

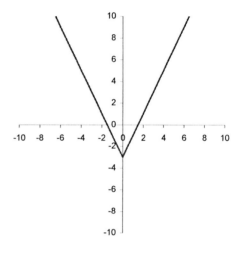

$y = 2|x| - 3$

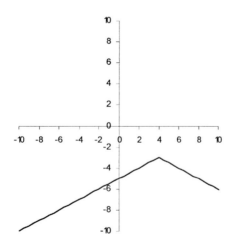

$y = {}^-1/2|x - 4| - 3$

-Note that on the first graph above, the graph opens up since m is positive 1. It has (⁻3,1) as its minimum point. The slopes of the 2 upward rays are ± 1.

- The second graph also opens up since m is positive. (0,⁻3) is its minimum point. The slopes of the 2 upward rays are ± 2.
- The third graph is a downward ∧ because m is ⁻1/2. The maximum point on the graph is at (4,⁻3). The slopes of the 2 downward rays are ± 1/2.

-The **identity function** is the linear equation $y = x$. Its graph is a line going through the origin (0,0) and through the first and third quadrants at a 45° degree angle.

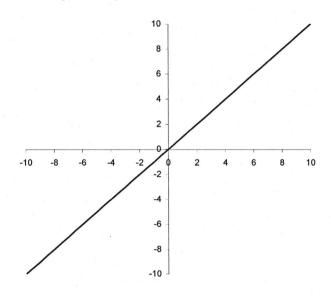

-The **greatest integer function** or **step function** has the equation: $f(x) = j[rx - h] + k$ or $y = j[rx - h] + k$. (h,k) is the location of the left endpoint of one step. j is the vertical jump from step to step. r is the reciprocal of the length of each step. If (x,y) is a point of the function, then when x is an integer, its y value is the same integer. If (x,y) is a point of the function, then when x is not an integer, its y value is the first integer less than x. Points on $y = [x]$ would include:

(3,3), (⁻2,⁻2), (0,0), (1.5,1), (2.83,2), (⁻3.2,⁻4), (⁻.4,⁻1).

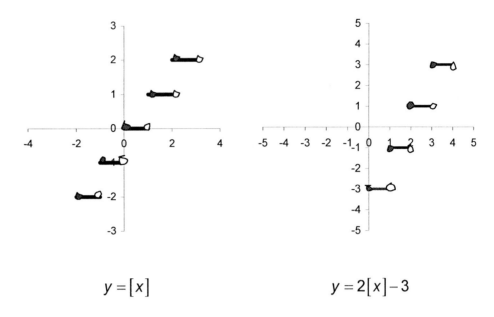

$$y = [x] \qquad\qquad y = 2[x] - 3$$

- Note that in the graph of the first equation, the steps are going up as they move to the right. Each step is one space wide (inverse of r) with a solid dot on the left and a hollow dot on the right where the jump to the next step occurs. Each step is one square higher ($j = 1$) than the previous step. One step of the graph starts at $(0,0) \leftarrow$ values of (h,k).

- In the second graph, the graph goes up to the right. One step starts at the point $(0, ^-3) \leftarrow$ values of (h,k). Each step is one square wide ($r = 1$) and each step is 2 squares higher than the previous step ($j = 2$).

Practice: Graph the following equations:

1. $f(x) = x$
2. $y = ^-|x - 3| + 5$
3. $y = 3[x]$
4. $y = 2/5|x - 5| - 2$

SKILL 8.4 Graph the solution set of first-degree equations or inequalities Involving one variable on a number line.

- When graphing a first-degree equation, solve for the variable. The graph of this solution will be a single point on the number line. There will be no arrows.

- When graphing a linear inequality, the dot will be hollow if the inequality sign is $<$ or $>$. If the inequality signs is either $\geq$ or $\leq$, the dot on the graph will be solid. The arrow goes to the right for $\geq$ or $>$. The arrow goes to the left for $\leq$ or $<$.

Solve:

$5(x+2)+2x=3(x-2)$

$5x+10+2x=3x-6$

$7x+10=3x-6$

$4x={^-}16$

$x={^-}4$

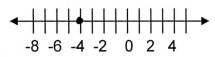

Solve:

$2(3x-7)>10x-2$

$6x-14>10x-2$

${^-}4x>12$

$x<{^-}3$ Note the change in inequality when dividint by negative numbers.

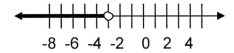

Solve the following equations and inequalities. Graph the solution set.

1. $5x-1>14$
2. $7(2x-3)+5x=19-x$
3. $3x+42\geq 12x-12$
4. $5-4(x+3)=9$

SKILL 8.5 Graph linear equations and inequalities involving two variables on the Cartesian plane.

- A first degree equation has an equation of the form $ax + by = c$. To graph this equation, find either one point and the slope of the line or find two points. To find a point and slope, solve the equation for y. This gets the equation in **slope intercept form**, $y = mx + b$. The point (0,b) is the y-intercept and m is the line's slope. To find any 2 points, substitute any 2 numbers for x and solve for y. To find the intercepts, substitute 0 for x and then 0 for y.

- Remember that graphs will go up as they go to the right when the slope is positive. Negative slopes make the lines go down as they go to the right.

- If the equation solves to **x = any number**, then the graph is a **vertical line**.

- If the equation solves to **y = any number**, then the graph is a **horizontal line**.

- When graphing a linear inequality, the line will be dotted if the inequality sign is < or >. If the inequality signs are either ≥ or ≤, the line on the graph will be a solid line. Shade above the line when the inequality sign is ≥ or >. Shade below the line when the inequality sign is < or ≤. Inequalities of the form $x >, x \leq, x <,$ or $x \geq$ number, draw a vertical line (solid or dotted). Shade to the right for > or ≥. Shade to the left for < or ≤. Remember: **Dividing or multiplying by a negative number will reverse the direction of the inequality sign.**

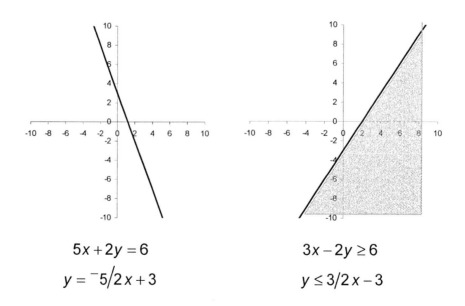

$5x + 2y = 6$ $3x - 2y \geq 6$

$y = {}^-5/2\,x + 3$ $y \leq 3/2\,x - 3$

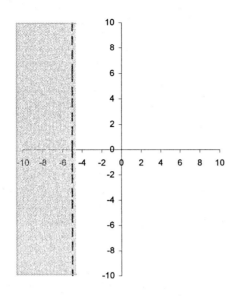

$$3x + 12 < -3$$
$$x < {}^-5$$

Graph the following:

1. $2x - y = {}^-4$
2. $x + 3y > 6$
3. $3x + 2y \leq 2y - 6$

TEACHER CERTIFICATION EXAM

SKILL 8.6 Identify the slope and intercepts of a graph or an equation.

- A first degree equation has an equation of the form $ax + by = c$. To find the slope of a line, solve the equation for y. This gets the equation into **slope intercept form**, $y = mx + b$. m is the line's slope.

- To find the y intercept, substitute 0 for x and solve for y. This is the y intercept. The y intercept is also the value of b in $y = mx + b$.

- To find the x intercept, substitute 0 for y and solve for x. This is the x intercept.

- If the equation solves to **x = any number**, then the graph is a **vertical line**. It only has an x intercept. Its slope is **undefined**.

- If the equation solves to **y = any number**, then the graph is a **horizontal line**. It only has a y intercept. Its slope is 0 (zero).

1. Find the slope and intercepts of $3x + 2y = 14$.

$$3x + 2y = 14$$
$$2y = {}^-3x + 14$$
$$y = {}^-3/2 \, x + 7$$

The slope of the line is $^-3/2$, the value of m.
The y intercept of the line is 7.

The intercepts can also be found by substituting 0 in place of the other variable in the equation.

To find the y intercept:	To find the x intercept:
let x = 0; 3(0) + 2y = 14	let y = 0; 3x + 2(0) = 14
0 + 2y = 14	3x + 0 = 14
2y = 14	3x = 14
y = 7	x = 14/3
(0,7) is the y intercept.	(14/3, 0) is the x intercept.

MATHEMATICS HIGH SCHOOL

Find the slope and the intercepts (if they exist) for these equations:

1. $5x + 7y = {}^-70$
2. $x - 2y = 14$
3. $5x + 3y = 3(5 + y)$
4. $2x + 5y = 15$

SKILL 8.7 Determine the equation of a line, given its graph.

- The equation of a graph can be found by finding its slope and its y intercept. To find the slope, find 2 points on the graph where co-ordinates are integer values. Using points: (x_1, y_1) and (x_2, y_2).

$$\text{slope} = \frac{y_2 - y_1}{x_2 - x_1}$$

The y intercept is the y coordinate of the point where a line crosses the y axis. The equation can be written in slope-intercept form, which is $y = mx + b$, where m is the slope and b is the y intercept. To rewrite the equation into some other form, multiply each term by the common denominator of all the fractions. Then rearrange terms as necessary.

- If the graph is a **vertical line**, then the equation solves to ***x* = the *x* co-ordinate of any point on the line**.

- If the graph is a **horizontal line**, then the equation solves to ***y* = the *y* coordinate of any point on the line**.

SKILL 8.8 Given two points, determine the equation of a line that contains the points.

- Given two points on a line, the first thing to do is to find the slope of the line. If 2 points on the graph are (x_1, y_1) and (x_2, y_2), then the slope is found using the formula:

$$\text{slope} = \frac{y_2 - y_1}{x_2 - x_1}$$

The slope will now be denoted by the letter **m**. To write the equation of a line, choose either point. Substitute them into the formula:

$$Y - y_a = m(X - x_a)$$

Remember (x_a, y_a) can be (x_1, y_1) or (x_2, y_2) If **m**, the value of the slope, is distributed through the parentheses, the equation can be rewritten into other forms of the equation of a line.

Find the equation of a line through $(9, {}^-6)$ and $({}^-1, 2)$.

$$\text{slope} = \frac{y_2 - y_1}{x_2 - x_1} = \frac{2 - {}^-6}{{}^-1 - 9} = \frac{8}{{}^-10} = \frac{{}^-4}{5}$$

$$Y - y_a = m(X - x_a) \rightarrow Y - 2 = {}^-4/5(X - {}^-1) \rightarrow$$
$$Y - 2 = {}^-4/5(X + 1) \rightarrow Y - 2 = {}^-4/5\, X - 4/5 \rightarrow$$
$$Y = {}^-4/5\, X + 6/5 \quad \text{This is the slope-intercept form.}$$

Multiplying by 5 to eliminate fractions, it is:

$$5Y = {}^-4X + 6 \rightarrow 4X + 5Y = 6 \quad \text{Standard form.}$$

Write the equation of a line through these two points:
1. $(5, 8)$ and $({}^-3, 2)$
2. $(11, 10)$ and $(11, {}^-3)$
3. $({}^-4, 6)$ and $(6, 12)$
4. $(7, 5)$ and $({}^-3, 5)$

TEACHER CERTIFICATION EXAM

COMPETENCY 9.0   **COMPREHENSION OF INTEGRAL EXPONENTS**

SKILL 9.1   Solve problems involving integral exponents.

Coefficients are the numbers in front of the variables in a term.

Addition/Subtraction with integral exponents.
-**Like terms** have the same variables raised to the same powers. Like terms can be combined by adding or subtracting their coefficients and by keeping the variables and their exponents the same. Unlike terms cannot be combined.

$$5x^4 + 3x^4 - 2x^4 = 6x^4$$
$$3ab^2 + 5ab + ab^2 = 4ab^2 + 5ab$$

Multiplication with integral exponents.
-When multiplying terms together, multiply their coefficients and add the exponents of the same variables. When a term is raised to a power, raise the term's coefficient to the power outside the parentheses. Then multiply the outside exponent times each of the inside exponents.

$$(3a^7b^5)(^-4a^2b^9c^4) = {}^-12a^9b^{14}c^4$$
$$(^-2x^3yz^7)^5 = {}^-32x^{15}y^5z^{35}$$

Division with integral exponents.
-When dividing any number of terms by a single term, divide or reduce their coefficients. Then subtract the exponent of a variable on the bottom from the exponent of the same variable on the top.

$$\frac{24a^8b^7 + 16a^7b^5 - 8a^6}{8a^6} = 3a^2b^7 + 2ab^5 - 1$$

Negative exponents are usually changed into positive exponents by moving those variables from numerator to denominator (or vice versa) to make the exponents become positive.

$$\frac{^-60a^8b^{-7}c^{-6}d^5e^{-12}}{12a^{-2}b^3c^{-9}d^5e^{-4}} = {}^-5a^{10}b^{-10}c^3e^{-8} = \frac{^-5a^{10}c^3}{b^{10}e^8}$$

$$(6x^{-9}y^3)(^-2x^5y^{-1}) = {}^-12x^{-4}y^2 = \frac{^-12y^2}{x^4}$$

MATHEMATICS HIGH SCHOOL

Simplify as far as possible:

1. $(4x^4y^2)(^-3xy^{-3}z^5)$
2. $(100a^6 + 60a^4 - 28a^2b^3) \div (4a^2)$
3. $(3x^5y^2)^4 + 8x^9y^6 + 19x^{20}y^8$

TEACHER CERTIFICATION EXAM

COMPETENCY 10.0   **ABILITY TO SOLVE PROBLEMS INVOLVING POLYNOMIALS AND RATIONAL ALGEBRAIC EXPRESSIONS.**

SKILL 10.1   Add or subtract rational algebraic fractions with like or unlike denominators.

- In order to add or subtract rational expressions, they must have a common denominator. If they don't have a common denominator, then factor the denominators to determine what factors are missing from each denominator to make the LCD. Multiply both numerator and denominator by the missing factor(s). Once the fractions have a common denominator, add or subtract their numerators, but keep the common denominator the same. Factor the numerator if possible and reduce if there are any factors that can be cancelled.

1. Find the least common denominator for $6a^3b^2$ and $4ab^3$.

     These factor into $2 \cdot 3 \cdot a^3 \cdot b^2$ and $2 \cdot 2 \cdot a \cdot b^3$.
     The first expression needs multiplied by another 2 and b.
     The other expression needs multiplied by 3 and a^2.
     Then both expressions would be $2 \cdot 2 \cdot 3 \cdot a^3 \cdot b^3 = 12a^3b^3 = \text{LCD}$.

2. Find the LCD for $x^2 - 4$, $x^2 + 5x + 6$, and $x^2 + x - 6$.

     $x^2 - 4$ factors into $(x-2)(x+2)$
     $x^2 + 5x + 6$ factors into $(x+3)(x+2)$
     $x^2 + x - 6$ factors into $(x+3)(x-2)$

To make these lists of factors the same, they must all be $(x+3)(x+2)(x-2)$. This is the LCD.

3.
$$\frac{5}{6a^3b^2} + \frac{1}{4ab^3} = \frac{5(2b)}{6a^3b^2(2b)} + \frac{1(3a^2)}{4ab^3(3a^2)} = \frac{10a}{12a^3b^3} + \frac{3b^2}{12a^3b^3} = \frac{10a + 3b^2}{12a^3b^3}$$

This will not reduce as all 3 terms are not divisible by anything.

MATHEMATICS HIGH SCHOOL

4.
$$\frac{2}{x^2-4} - \frac{3}{x^2+5x+6} + \frac{7}{x^2+x-6} =$$

$$\frac{2}{(x-2)(x+2)} - \frac{3}{(x+3)(x+2)} + \frac{7}{(x+3)(x-2)} =$$

$$\frac{2(x+3)}{(x-2)(x+2)(x+3)} - \frac{3(x-2)}{(x+3)(x+2)(x-2)} + \frac{7(x+2)}{(x+3)(x-2)(x+2)} =$$

$$\frac{2x+6}{(x-2)(x+2)(x+3)} - \frac{3x-6}{(x+3)(x+2)(x-2)} + \frac{7x+14}{(x+3)(x-2)(x+2)} =$$

$$\frac{2x+6-(3x-6)+7x+14}{(x+3)(x-2)(x+2)} = \frac{6x+26}{(x+3)(x-2)(x+2)}$$

This will not reduce.

Try These:

5. $\dfrac{6}{x-3} + \dfrac{2}{x+7}$

6. $\dfrac{5}{4a^2b^5} + \dfrac{3}{5a^4b^3}$

7. $\dfrac{x+3}{x^2-25} + \dfrac{x-6}{x^2-2x-15}$

TEACHER CERTIFICATION EXAM

SKILL 10.2 Solve word problems involving the use of rational algebraic expressions and equations.

- Some problems can be solved using equations with rational expressions. First write the equation. To solve it, multiply each term by the LCD of all fractions. This will cancel out all of the denominators and give an equivalent algebraic equation that can be solved.

1. The denominator of a fraction is two less than three times the numerator. If 3 is added to both the numerator and denominator, the new fraction equals 1/2.

original fraction: $\dfrac{x}{3x-2}$ revised fraction: $\dfrac{x+3}{3x+1}$

$$\dfrac{x+3}{3x+1} = \dfrac{1}{2} \qquad 2x+6 = 3x+1$$
$$x = 5$$

original fraction: $\dfrac{5}{13}$

2. Elly Mae can feed the animals in 15 minutes. Jethro can feed them in 10 minutes. How long will it take them if they work together?

Solution: If Elly Mae can feed the animals in 15 minutes, then she could feed 1/15 of them in 1 minute, 2/15 of them in 2 minutes, $x/15$ of them in x minutes. In the same fashion Jethro could feed $x/10$ of them in x minutes. Together they complete 1 job. The equation is:

$$\dfrac{x}{15} + \dfrac{x}{10} = 1$$

Multiply each term by the LCD of 30:

$$2x + 3x = 30$$
$$x = 6 \text{ minutes}$$

MATHEMATICS HIGH SCHOOL

TEACHER CERTIFICATION EXAM

3. A salesman drove 480 miles from Pittsburgh to Hartford. The next day he returned the same distance to Pittsburgh in half an hour less time than his original trip took, because he increased his average speed by 4 mph. Find his original speed.

Since distance = rate x time then time = $\dfrac{\text{distance}}{\text{rate}}$

original time $-1/2$ hour $=$ shorter return time

$$\frac{480}{x} - \frac{1}{2} = \frac{480}{x+4}$$

Multiplying by the LCD of $2x(x+4)$, the equation becomes:

$$480\big[2(x+4)\big] - 1\big[x(x+4)\big] = 480(2x)$$
$$960x + 3840 - x^2 - 4x = 960x$$
$$x^2 + 4x - 3840 = 0$$
$$(x+64)(x-60) = 0$$
$$x = 60 \qquad \text{60 mph is the original speed}$$
$$\qquad\qquad\quad \text{64 mph is the faster return speed}$$

Try these:

1. Working together, Larry, Moe, and Curly can paint an elephant in 3 minutes. Working alone, it would take Larry 10 minutes or Moe 6 minutes to paint the elephant. How long would it take Curly to paint the elephant if he worked alone?

2. The denominator of a fraction is 5 more than twice the numerator. If the numerator is doubled, and the denominator is increased by 5, the new fraction is equal to $1/2$. Find the original number.

3. A trip from Augusta, Maine to Galveston, Texas is 2108 miles. If one car drove 6 mph faster than a truck and got to Galveston 3 hours before the truck, find the speeds of the car and truck.

MATHEMATICS HIGH SCHOOL

TEACHER CERTIFICATION EXAM

SKILL 10.3 Solve algebraic formulas for one particular variable.

- To solve an algebraic formula for some variable, called R, follow the following steps:

a. Eliminate any parentheses using the distributive property.
b. Multiply every term by the LCD of any fractions to write an equivalent equation without any fractions.
c. Move all terms containing the variable, R, to one side of the equation. Move all terms without the variable to the opposite side of the equation.
d. If there are 2 or more terms containing the variable R, factor **only R** out of each of those terms as a common factor.
e. Divide both sides of the equation by the number or expression being multiplied times the variable, R.
f. Reduce fractions if possible.
g. Remember there are restrictions on values allowed for variables because the denominator can not equal zero.

1. Solve $A = p + prt$ for t.

$$A - p = prt$$
$$\frac{A-p}{pr} = \frac{prt}{pr}$$
$$\frac{A-p}{pr} = t$$

2. Solve $A = p + prt$ for p.

$$A = p(1+rt)$$
$$\frac{A}{1+rt} = \frac{p(1+rt)}{1+rt}$$
$$\frac{A}{1+rt} = p$$

3. $A = 1/2\, h(b_1 + b_2)$ for b_2

$$A = 1/2\, hb_1 + 1/2\, hb_2 \quad \leftarrow \text{step a}$$
$$2A = hb_1 + hb_2 \quad \leftarrow \text{step b}$$
$$2A - hb_1 = hb_2 \quad \leftarrow \text{step c}$$
$$\frac{2A - hb_1}{h} = \frac{hb_2}{h} \quad \leftarrow \text{step d}$$
$$\frac{2A - hb_1}{h} = b_2 \quad \leftarrow \text{will not reduce}$$

Solve:
1. $F = 9/5\, C + 32$ for C
2. $A = 1/2\, bh + h^2$ for b
3. $S = 180(n - 2)$ for n

MATHEMATICS HIGH SCHOOL

TEACHER CERTIFICATION EXAM

COMPETENCY 11.0 UNDERSTANDING POLYNOMIAL FACTORIZATION OVER THE RATIONAL NUMBERS.

SKILL 11.1 Factor the sum or difference of two cubes.

- To factor the sum or the difference of perfect cubes, follow this procedure:

a. Factor out any greatest common factor (GCF).

b. Make a parentheses for a binomial (2 terms) followed by a trinomial (3 terms).

c. The sign in the first parentheses is the same as the sign in the problem. The difference of cubes will have a "-" sign in the first parentheses. The sum of cubes will use a "+".

d. The first sign in the second parentheses is the opposite of the sign in the first parentheses. The second sign in the other parentheses is always a "+".

e. Determine what would be cubed to equal each term of the problem. Put those expressions in the first parentheses.

f. To make the 3 terms of the trinomial, think square - product - square. Looking at the binomial, square the first term. This is the trinomial's first term. Looking at the binomial, find the product of the two terms, ignoring the signs. This is the trinomial's second term. Looking at the binomial, square the third term. This is the trinomial's third term. Except in rare instances, the trinomial does not factor again.

Factor completely:

1.
$16x^3 + 54y^3$

$2(8x^3 + 27y^3)$ ← GCF

$2(\quad + \quad)(\quad - \quad + \quad)$ ← signs

$2(2x + 3y)(\quad - \quad + \quad)$ ← what is cubed to equal $8x^3$ or $27y^3$

$2(2x + 3y)(4x^2 - 6xy + 9y^2)$ ← square-product-square

2.

$64a^3 - 125b^3$

$(\quad - \quad)(\quad + \quad + \quad)$ ← signs

$(4a - 5b)(\quad + \quad + \quad)$ ← what is cubed to equal $64a^3$ or $125b^3$

$(4a - 5b)(16a^2 + 20ab + 25b^2)$ ← square-product-square

3.

$27x^{27} + 343y^{12} = (3x^9 + 7y^{12})(9x^{18} - 21x^9 y^{12} + 49y^{24})$

Note: The coefficient 27 is different from the exponent 27.

Try These:

1. $216x^3 - 125y^3$
2. $4a^3 - 32b^3$
3. $40x^{29} + 135x^2 y^3$

TEACHER CERTIFICATION EXAM

SKILL 11.2 Factor polynomials completely.

- To factor a polynomial, follow these steps:

a. **Factor out any GCF** (greatest common factor)

b. For a binomial (2 terms), check to see if the problem is the **difference of perfect squares**. If both factors are perfect squares, then it factors this way:
$$a^2 - b^2 = (a-b)(a+b)$$

If the problem is not the difference of perfect squares, then check to see if the problem is either the sum or difference of perfect cubes.

$$x^3 - 8y^3 = (x-y)(x^2 + xy + y^2) \quad \leftarrow \text{difference}$$

$$64a^3 + 27b^3 = (4a+3b)(16a^3 - 12ab + 9b^2) \quad \leftarrow \text{sum}$$

** The sum of perfect squares does NOT factor.

c. Trinomials could be perfect squares. Trinomials can be factored into 2 binomials (un-FOILing). Be sure the terms of the trinomial are in descending order. If last sign of the trinomial is a "+", then the signs in the parentheses will be the same as the sign in front of the second term of the trinomial. If the last sign of the trinomial is a "-", then there will be one "+" and one "-" in the two parentheses. The first term of the trinomial can be factored to equal the first terms of the two factors. The last term of the trinomial can be factored to equal the last terms of the two factors. Work backwards to determine the correct factors to multiply together to get the correct center term.

Factor completely:

1. $4x^2 - 25y^2$
2. $6b^2 - 2b - 8$
3. Find a factor of $6x^2 - 5x - 4$

    a. $(3x+2)$ b. $(3x-2)$ c. $(6x-1)$ d. $(2x+1)$

Answers:

1. No GCF; this is the difference of perfect squares.

$$4x^2 - 25y^2 = (2x - 5y)(2x + 5y)$$

2. GCF of 2; Try to factor into 2 binomials:

$$6b^2 - 2b - 8 = 2(3b^2 - b - 4)$$

Signs are one "+", one "−". $3b^2$ factors into $3b$ and b. Find factors of 4: 1 & 4; 2 & 2.

$$6b^2 - 2b - 8 = 2(3b^2 - b - 4) = 2(3b - 4)(b + 1)$$

3. If an answer choice is correct, find the other factor:

   a. $(3x + 2)(2x - 2) = 6x^2 - 2x - 4$
   b. $(3x - 2)(2x + 2) = 6x^2 + 2x - 4$
   c. $(6x - 1)(x + 4) = 6x^2 + 23x - 4$
   d. $(2x + 1)(3x - 4) = 6x^2 - 5x - 4$ ← correct factors

TEACHER CERTIFICATION EXAM

COMPETENCY 12.0 **MASTERY OF RATIONAL AND IRRATIONAL EXPRESSIONS.**

SKILL 12.1 Determine equivalent forms of given rational expressions.

Rational expressions can be changed into other equivalent fractions by either reducing them or by changing them to have a common denominator. When dividing any number of terms by a single term, divide or reduce their coefficients. Then subtract the exponent of a variable on the bottom from the exponent of the same variable from the numerator.

To reduce a rational expression with more than one term in the denominator, the expression must be factored first. Factors that are exactly the same will cancel and each become a 1. Factors that have exactly the opposite signs of each other, such as $(a-b)$ and $(b-a)$, will cancel and one factor becomes a 1 and the other becomes a $^-1$.

To make a fraction have a common denominator, factor the fraction. Determine what factors are missing from that particular denominator, and multiply both the numerator and the denominator by those missing factors. This gives a new fraction which now has the common denominator.

Simplify these fractions:

1. $\dfrac{24x^3y^6z^3}{8x^2y^2z} = 3xy^4z^2$

2. $\dfrac{3x^2-14xy-5y^2}{x^2-25y^2} = \dfrac{(3x+y)(x-5y)}{(x+5y)(x-5y)} = \dfrac{3x+y}{x+5y}$

3. Re-write this fraction with a denominator of $(x+3)(x-5)(x+4)$.

$\dfrac{x+2}{x^2+7x+12} = \dfrac{x+2}{(x+3)(x+4)} = \dfrac{(x+2)(x-5)}{(x+3)(x+4)(x-5)}$

TEACHER CERTIFICATION EXAM

Try these:

1. $\dfrac{72x^4 y^9 z^{10}}{8x^3 y^9 z^5}$

2. $\dfrac{3x^2 - 13xy - 10y^2}{x^3 - 125y^3}$

3. Re-write this fraction with a denominator of $(x+2)(x+3)(x-7)$.

$$\dfrac{x+5}{x^2 - 5x - 14}$$

SKILL 12.2 Perform the four basic operations with rational expressions.

In order **to add or subtract** rational expressions, they must have a common denominator. If they don't have a common denominator, then factor the denominators to determine what factors are missing from each denominator to make the LCD. Multiply both numerator and denominator by the missing factor(s). Once the fractions have a common denominator, add or subtract their numerators, but keep the common denominator the same. Factor the numerator if possible and reduce if there are any factors that can be cancelled.

In order **to multiply** rational expressions, they do not have to have a common denominator. If you factor each numerator and denominator, you can cancel out any factor that occurs in both the numerator and denominator. Then multiply the remaining factors of the numerator together. Last multiply the remaining factors of the denominator together.

In order **to divide** rational expressions, the problem must be re-written as the first fraction multiplied times the inverse of the second fraction. Once the problem has been written as a multiplication, factor each numerator and denominator. Cancel out any factor that occurs in both the numerator and denominator. Then multiply the remaining factors of the numerator together. Last multiply the remaining factors of the denominator together.

1. $\dfrac{5}{x^2-9} - \dfrac{2}{x^2+4x+3} = \dfrac{5}{(x-3)(x+3)} - \dfrac{2}{(x+3)(x+1)} =$

$\dfrac{5(x+1)}{(x+1)(x-3)(x+3)} - \dfrac{2(x-3)}{(x+3)(x+1)(x-3)} = \dfrac{3x+11}{(x-3)(x+3)(x+1)}$

2. $\dfrac{x^2-2x-24}{x^2+6x+8} \times \dfrac{x^2+3x+2}{x^2-13x+42} = \dfrac{(x-6)(x+4)}{(x+4)(x+2)} \times \dfrac{(x+2)(x+1)}{(x-7)(x-6)} = \dfrac{x+1}{x-7}$

Try these:

1. $\dfrac{6}{x^2-1} + \dfrac{8}{x^2+7x+6}$

2. $\dfrac{x^2-9}{x^2-4} \div \dfrac{x^2+8x+15}{x^3+8}$

SKILL 12.3 Solve equations containing radical expressions.

To solve an equation with rational expressions, find the least common denominator of all the fractions. Multiply each term by the LCD of all fractions. This will cancel out all of the denominators and give an equivalent algebraic equation that can be solved. Solve the resulting equation. Once you have found the answer(s), substitute them back into the original equation to check them. Sometimes there are solutions that do not check in the original equation. These are extraneous solutions, which are not correct and must be eliminated. If a problem has more than one potential solution, each solution must be checked separately.

> **NOTE: What this really means is that you can substitute the answers from any multiple choice test back into the question to determine which answer choice is correct.**

TEACHER CERTIFICATION EXAM

Solve and **check**:

1. $\dfrac{72}{x+3} = \dfrac{32}{x+3} + 5$ 　　　LCD $= x + 3$, so multiply by this.

$(x+3) \times \dfrac{72}{x+3} = (x+3) \times \dfrac{32}{x+3} + 5(x+3)$

$72 = 32 + 5(x+3) \rightarrow 72 = 32 + 5x + 15$

$72 = 47 + 5x \qquad \rightarrow 25 = 5x$

$5 = x$ (This checks too).

2. $\dfrac{12}{2x^2 - 4x} + \dfrac{13}{5} = \dfrac{9}{x-2}$ 　　Factor $2x^2 - 4x = 2x(x-2)$.

　　　　　　　　　　　　　　　LCD $= 5 \times 2x(x-2)$ or $10x(x-2)$

$10x(x-2) \times \dfrac{12}{2x(x-2)} + 10x(x-2) \times \dfrac{13}{5} = \dfrac{9}{x-2} \times 10x(x-2)$

$60 + 2x(x-2)(13) = 90x$

$26x^2 - 142x + 60 = 0$

$2(13x^2 - 71x + 30) = 0$

$2(x-5)(13x-6)$ 　　so $x = 5$ or $x = 6/13$ 　← both check

Try these:

1. $\dfrac{x+5}{3x-5} + \dfrac{x-3}{2x+2} = 1$

2. $\dfrac{2x-7}{2x+5} = \dfrac{x-6}{x+8}$

MATHEMATICS HIGH SCHOOL

TEACHER CERTIFICATION EXAM

SKILL 12.4 Simplify radical expressions.

To simplify a radical, follow these steps:

First factor the number or coefficient completely.

- For square roots, group like-factors in groups of 2. For cube roots, group like-factors in groups of 3. For n^{th} roots, group like-factors in groups of n.
- Now, for each of those groups, put one of that number outside the radical. Multiply   these numbers by any number already in front of the radical. Any factors that were not combined in groups should be multiplied back together and left inside the radical.
- The index number of a radical is the little number on the front of the radical. For a cube root, the index is a 3. If no index appears, then the index is a 2 (for square roots).
- For variables inside the radical, divide the index number of the radical into each exponent. The quotient (the answer to the division) is the new exponent to be written on the variable outside the radical. The remainder from the division is the new exponent on the variable remaining inside the radical sign. If the remainder is zero, then the variable no longer appears inside the radical sign.
- Note: Remember that the square root of a negative number can be done by replacing the negative sign inside the square root sign with an "i" in front of the radical (to indicate an imaginary number). Then simplify the remaining positive radical by the normal method. Include the "i" outside the radical as part of the answer.

$$\sqrt{-18} = i\sqrt{18} = i\sqrt{3 \cdot 3 \cdot 2} = 3i\sqrt{2}$$

- Remember that if the index number is an odd number, you can still simplify the radical to get a negative solution.

Simplify:

1. $\sqrt{50a^4b^7} = \sqrt{5 \cdot 5 \cdot 2 \cdot a^4 b^7} = 5a^2 b^3 \sqrt{2b}$
2. $7x \sqrt[3]{16x^5} = 7x \sqrt[3]{2 \cdot 2 \cdot 2 \cdot 2 \cdot x^5} = 7x \cdot 2x \sqrt[3]{2x^2} = 14x^2 \sqrt[3]{2x^2}$

Try These :

1. $\sqrt{72a^9}$
2. $\sqrt{-98}$
3. $\sqrt[3]{-8x^6}$
4. $2x^3 y \sqrt[4]{243x^6 y^{11}}$

MATHEMATICS HIGH SCHOOL

TEACHER CERTIFICATION EXAM

SKILL 12.5 Multiply or divide binomials containing radicals.

The conjugate of a binomial is the same expression as the original binomial with the sign between the 2 terms changed.

$$\text{The conjugate of } 3+2\sqrt{5} \text{ is } 3-2\sqrt{5}.$$
$$\text{The conjugate of } \sqrt{5}-\sqrt{7} \text{ is } \sqrt{5}+\sqrt{7}.$$
$$\text{The conjugate of } {}^-6-\sqrt{11} \text{ is } {}^-6+\sqrt{11}.$$

To multiply binomials including radicals, "FOIL" the binomials together. (that is, distribute each term of the first binomial times each term of the second binomial). Multiply what is in front of the radicals together. Multiply what is inside of the two radicals together. Check to see if any of the radicals can be simplified. Combine like terms, if possible.

When one binomial is divided by another binomial, multiply both the numerator and denominator by the conjugate of the denominator. "FOIL" or distribute one binomial through the other binomial. Simplify the radicals, if possible, and combine like terms. Reduce the resulting fraction if every term is divisible outside the radical signs by the same number.

1. $(5+\sqrt{10})(4-3\sqrt{2}) = 20-15\sqrt{2}+4\sqrt{10}-3\sqrt{20} =$
   $20-15\sqrt{2}+4\sqrt{10}-6\sqrt{5}$

2. $(\sqrt{6}+5\sqrt{2})(3\sqrt{6}-8\sqrt{2}) = 3\sqrt{36}-8\sqrt{12}+15\sqrt{12}-40\sqrt{4} =$
   $3\cdot 6 - 8\cdot 2\sqrt{3} + 15\cdot 2\sqrt{3} - 40\cdot 2 = 18 - 16\sqrt{3} + 30\sqrt{3} - 80 = {}^-62 + 14\sqrt{3}$

3. $\dfrac{1-\sqrt{2}}{3+5\sqrt{2}} = \dfrac{1-\sqrt{2}}{3+5\sqrt{2}} \cdot \dfrac{3-5\sqrt{2}}{3-5\sqrt{2}} = \dfrac{3-5\sqrt{2}-3\sqrt{2}+5\sqrt{4}}{9-25\sqrt{4}} = \dfrac{3-5\sqrt{2}-3\sqrt{2}+10}{9-50} =$
   $\dfrac{13-8\sqrt{2}}{{}^-41}$ or $-\dfrac{13-8\sqrt{2}}{41}$ or $\dfrac{{}^-13+8\sqrt{2}}{41}$

Try These:
1. $(3+2\sqrt{6})(4-\sqrt{6})$
2. $(\sqrt{5}+2\sqrt{15})(\sqrt{3}-\sqrt{15})$
3. $\dfrac{6+2\sqrt{3}}{4-\sqrt{6}}$

MATHEMATICS HIGH SCHOOL

TEACHER CERTIFICATION EXAM

SKILL 12.6 Add, subtract, multiply, divide, and simplify radical expressions (limited to square roots).

Before you can add or subtract square roots, the numbers or expressions inside the radicals must be the same. First, simplify the radicals, if possible. If the numbers or expressions inside the radicals are the same, add or subtract the numbers (or like expressions) in front of the radicals. Keep the expression inside the radical the same. Be sure that the radicals are as simplified as possible.

Note: If the expressions inside the radicals are not the same, and can not be simplified to become the same, then they can not be combined by addition or subtraction.

To multiply 2 square roots together, follow these steps:

    1. Multiply what is outside the radicals together.

    2. Multiply what is inside the radicals together.

    3. Simplify the radical if possible. Multiply whatever is in front of the radical times the expression that is coming out of the radical.

To divide one square root by another, follow these steps:

    1. Work separately on what is inside or outside the square root sign.

    2. Divide or reduce the coefficients outside the radical.

    3. Divide any like variables outside the radical.

    4. Divide or reduce the coefficients inside the radical.

    5. Divide any like variables inside the radical.

    6. If there is still a radical in the denominator, multiply both the numerator and denominator by the radical in the denominator. Simplify both resulting radicals and reduce again outside the radical (if possible).

MATHEMATICS HIGH SCHOOL

Simplify:

1. $6\sqrt{7} + 2\sqrt{5} + 3\sqrt{7} = 9\sqrt{7} + 2\sqrt{5}$ These cannot be combined further.

2. $5\sqrt{12} + \sqrt{48} - 2\sqrt{75} = 5\sqrt{2 \cdot 2 \cdot 3} + \sqrt{2 \cdot 2 \cdot 2 \cdot 2 \cdot 3} - 2\sqrt{3 \cdot 5 \cdot 5} =$
$5 \cdot 2\sqrt{3} + 2 \cdot 2\sqrt{3} - 2 \cdot 5\sqrt{3} = 10\sqrt{3} + 4\sqrt{3} - 10\sqrt{3} = 4\sqrt{3}$

3. $(6\sqrt{15x})(7\sqrt{10x}) = 42\sqrt{150x^2} = 42\sqrt{2 \cdot 3 \cdot 5 \cdot 5 \cdot x^2} = 42 \cdot 5x\sqrt{2 \cdot 3} = 210x\sqrt{6}$

4. $\dfrac{105x^8 \sqrt{18x^5y^6}}{30x^2 \sqrt{27x^2y^4}} = \dfrac{7x^{10}y^3 \sqrt{2x}}{2x^3y^2\sqrt{3}} = \dfrac{7x^7y\sqrt{2x}}{2\sqrt{3}} \times \dfrac{\sqrt{3}}{\sqrt{3}} = \dfrac{7x^7y\sqrt{6x}}{2\sqrt{9}} =$
$\dfrac{7x^7y\sqrt{6x}}{6}$

Try these:

1. $6\sqrt{24} + 3\sqrt{54} - \sqrt{96}$
2. $\left(2x^2y\sqrt{18x}\right)\left(7xy^7 \sqrt{4x}\right)$
3. $\dfrac{125a^5 \sqrt{56a^4b^7}}{40a^2 \sqrt{40a^2b^8}}$
4. $2\sqrt{3} + 4\sqrt{5} + 6\sqrt{25} - 7\sqrt{9} + 2\sqrt{5} - 8\sqrt{20} - 6\sqrt{16} - 7\sqrt{3}$

SKILL 12.7 Solve equations involving radicals.

To solve a radical equation, follow these steps:

1. Get a radical alone on one side of the equation.
2. Raise both **sides** of the equation to the power equal to the index number. **Do not raise them to that power term by term, but raise the entire side to that power**. Combine any like terms.
3. If there is another radical still in the equation, repeat steps one and two (i.e. get that radical alone on one side of the equation and raise both sides to a power equal to the index). Repeat as necessary until the radicals are all gone.

TEACHER CERTIFICATION EXAM

4. Solve the resulting equation.
5. Once you have found the answer(s), substitute them back into the original equation to check them. Sometimes there are solutions that do not check in the original equation. These are extraneous solutions, which are not correct and must be eliminated. If a problem has more than one potential solution, each solution must be checked separately.

NOTE: What this really means is that you can substitute the answers from any multiple choice test back into the question to determine which answer choice is correct.

Solve and **check**.

1. $\sqrt{2x+1} + 7 = x$

   $\sqrt{2x+1} = x - 7$

   $\left(\sqrt{2x+1}\right)^2 = (x-7)^2$ ← BOTH sides are squared.

   $2x + 1 = x^2 - 14x + 49$

   $0 = x^2 - 16x + 48$

   $0 = (x-12)(x-4)$

   $x = 12, \ x = 4$

When you check these answers in the original equation, 12 checks; however, **4 does not check in the original equation**. Therefore, the only answer is x = 12.

2. $\sqrt{3x+4} = 2\sqrt{x-4}$

   $\left(\sqrt{3x+4}\right)^2 = \left(2\sqrt{x-4}\right)^2$

   $3x + 4 = 4(x-4)$

   $3x + 4 = 4x - 16$

   $20 = x$ ← This checks in the original equaion.

3. $\sqrt[4]{7x-3} = 3$

   $\left(\sqrt[4]{7x-3}\right)^4 = 3^4$

   $7x - 3 = 81$

   $7x = 84$

   $x = 12$ ← This checks out with the original equation.

4. $\sqrt{x} = {}^-3$

$\left(\sqrt{x}\right)^2 = \left({}^-3\right)^2$

$x = 9$ ← This does NOT check in the original equation. Since there is no other answer to check, the correct answer is the empty set or the null set or $\emptyset$.

Try these:

Solve and check.

1. $\sqrt{8x - 24} + 14 = 2x$
2. $\sqrt{6x - 17} = 3\sqrt{x + 2} - 4$

   Hint: Remember to square both sides, not term by term!
3. $\sqrt{6x - 2} = 5\sqrt{x - 13}$

TEACHER CERTIFICATION EXAM

COMPETENCY 13.0   KNOWLEDGE OF QUADRATIC EQUATIONS AND INEQUALITIES

SKILL 13.1 Solve quadratic equations by factoring, graphing, completing the square, or using the quadratic formula.

A **quadratic equation** is written in the form $ax^2 + bx + c = 0$. To solve a quadratic equation by factoring, at least one of the factors must equal zero.

Example:
Solve the equation.

$x^2 + 10x - 24 = 0$
$(x + 12)(x - 2) = 0$ Factor.
$x + 12 = 0$ or $x - 2 = 0$ Set each factor equal to 0.
$x = {}^-12$ $x = 2$ Solve.

Check:
$x^2 + 10x - 24 = 0$
$({}^-12)^2 + 10({}^-12) - 24 = 0$ $(2)^2 + 10(2) - 24 = 0$
$144 - 120 - 24 = 0$ $4 + 20 - 24 = 0$
$0 = 0$ $0 = 0$

A quadratic equation that cannot be solved by factoring can be solved by **completing the square**.

Example:

Solve the equation.

$x^2 - 6x + 8 = 0$
$x^2 - 6x = {}^-8$ Move the constant to the right side.
$x^2 - 6x + 9 = {}^-8 + 9$ Add the square of half the cooeffient of x to both sides.
$(x - 3)^2 = 1$ Write the left side as a perfect square.
$x - 3 = \pm\sqrt{1}$ Take the square root of both sides.
$x - 3 = 1$ $x - 3 = {}^-1$ Solve.
$x = 4$ $x = 2$

MATHEMATICS HIGH SCHOOL

Check:
$x^2 - 6x + 8 = 0$

$4^2 - 6(4) + 8 = 0$ $\qquad$ $2^2 - 6(2) + 8 = 0$
$16 - 24 + 8 = 0$ $\qquad$ $4 - 12 + 8 = 0$
$0 = 0$ $\qquad\qquad\qquad$ $0 = 0$

The general technique for graphing quadratics is the same as for graphing linear equations. Graphing quadratic equations, however, results in a parabola instead of a straight line.

Example:

Graph $y = 3x^2 + x - 2$.

x	$y = 3x^2 + x - 2$
-2	8
-1	0
0	-2
1	2
2	12

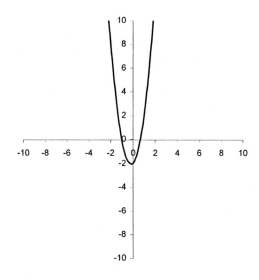

To solve a quadratic equation using the quadratic formula, be sure that your equation is in the form $ax^2 + bx + c = 0$. Substitute these values into the formula:

$$\frac{-b \pm \sqrt{b^2 - 4ac}}{2a}$$

Example:

Solve the equation.

$3x^2 = 7 + 2x \rightarrow 3x^2 - 2x - 7 = 0$

$a = 3 \quad b = {}^-2 \quad c = {}^-7$

$x = \dfrac{-({}^-2) \pm \sqrt{({}^-2)^2 - 4(3)({}^-7)}}{2(3)}$

$x = \dfrac{2 \pm \sqrt{4 + 84}}{6}$

$x = \dfrac{2 \pm \sqrt{88}}{6}$

$x = \dfrac{2 \pm 2\sqrt{22}}{6}$

$x = \dfrac{1 \pm \sqrt{22}}{3}$

SKILL 13.2 Use quadratic equations to solve word problems.

Some word problems will give a quadratic equation to be solved. When the quadratic equation is found, set it equal to zero and solve the equation by factoring or the quadratic formula. Examples of this type of problem follow.

Example:
Alberta (A) is a certain distance north of Boston (B). The distance from Boston east to Carlisle (C) is 5 miles more than the distance from Alberta to Boston. The distance from Alberta to Carlisle is 10 miles more than the distance from Alberta to Boston. How far is Alberta from Carlisle?

Solution:
Since north and east form a right angle, these distances are the lengths of the legs of a right triangle. If the distance from Alberta to Boston is x, then from Boston to Carlisle is $x+5$, and the distance from Alberta to Carlisle is $x+10$.

The equation is: $AB^2 + BC^2 = AC^2$

$$x^2 + (x+5)^2 = (x+10)^2$$
$$x^2 + x^2 + 10x + 25 = x^2 + 20x + 100$$
$$2x^2 + 10x + 25 = x^2 + 20x + 100$$
$$x^2 - 10x - 75 = 0$$
$$(x-15)(x+5) = 0 \quad \text{Distance cannot be negative.}$$
$$x = 15 \quad \text{Distance from Alberta to Boston.}$$
$$x + 5 = 20 \quad \text{Distance from Boston to Carlisle.}$$
$$x + 10 = 25 \quad \text{Distance from Alberta to Carlisle.}$$

Example:
The square of a number is equal to 6 more than the original number. Find the original number.

Solution: If x = original number, then the equation is:

$$x^2 = 6 + x \quad \text{Set this equal to zero.}$$
$$x^2 - x - 6 = 0 \quad \text{Now factor.}$$
$$(x-3)(x+2) = 0$$
$$x = 3 \text{ or } x = {}^-2 \quad \text{There are 2 solutions, 3 or } {}^-2.$$

Try these:

1. One side of a right triangle is 1 less than twice the shortest side, while the third side of the triangle is 1 more than twice the shortest side. Find all 3 sides.

2. Twice the square of a number equals 2 less than 5 times the number. Find the number(s).

SKILL 13.3 Use the discriminant to determine the nature of solids.

The discriminant of a quadratic equation is the part of the quadratic formula that is usually inside the radical sign, $b^2 - 4ac$.

$$x = \frac{-b \pm \sqrt{b^2 - 4ac}}{2a}$$

The radical sign is NOT part of the discriminant!! Determine the value of the discriminant by substituting the values of a, b, and c from $ax^2 + bx + c = 0$.

-If the value of the discriminant is **any negative number**, then there are **two complex roots** including "i".
-If the value of the discriminant is **zero**, then there is only **1 real rational root**. This would be a double root.
-If the value of the discriminant is **any positive number that is also a perfect square**, then there are **two real rational roots.** (There are no longer any radical signs.)
-If the value of the discriminant is **any positive number that is NOT a perfect square**, then there are **two real irrational roots.** (There are still unsimplified radical signs.)

Example:

Find the value of the discriminant for the following equations. Then determine the number and nature of the solutions of that quadratic equation.

$2x^2 - 5x + 6 = 0$
a = 2, b = ⁻5, c = 6 so $b^2 - 4ac = (^-5)^2 - 4(2)(6) = 25 - 48 = {^-}23$.

Since ⁻23 is a negative number, there are **two complex roots** including "i".

$3x^2 - 12x + 12 = 0$
a = 3, b = ⁻12, c = 12 so $b^2 - 4ac = (^-12)^2 - 4(3)(12) = 144 - 144 = 0$.

Since 0 is the value of the discriminant, there is only
1 real rational root.

$6x^2 - x - 2 = 0$
a = 6, b = ⁻1, c = ⁻2 so $b^2 - 4ac = (^-1)^2 - 4(6)(^-2) = 1 + 48 = 49$.

Since 49 is positive and is also a perfect square $(\sqrt{49}) = 7$, then there are **two real rational roots.**

Try these:

1. $6x^2 - 7x - 8 = 0$
2. $10x^2 - x - 2 = 0$
3. $25x^2 - 80x + 64 = 0$

TEACHER CERTIFICATION EXAM

SKILL 13.4 Determine a quadratic from known roots.

Follow these steps to write a quadratic equation from its roots:

1. Add the roots together. The answer is their **sum**. Multiply the roots together. The answer is their **product**.
2. A quadratic equation can be written using the sum and product like this:

$$x^2 + (\text{opposite of the sum})x + \text{product} = 0$$

3. If there are any fractions in the equation, multiply every term by the common denominator to eliminate the fractions. This is the quadratic equation.
4. If a quadratic equation has only 1 root, use it twice and follow the first 3 steps above.

Example:
Find a quadratic equation with roots of 4 and $^-9$.

Solutions:
The sum of 4 and $^-9$ is $^-5$. The product of 4 and $^-9$ is $^-36$.
The equation would be:

$$x^2 + (\text{opposite of the sum})x + \text{product} = 0$$
$$x^2 + 5x - 36 = 0$$

Find a quadratic equation with roots of $5 + 2i$ and $5 - 2i$.

Solutions:
The sum of $5 + 2i$ and $5 - 2i$ is 10. The product of $5 + 2i$ and $5 - 2i$ is $25 - 4i^2 = 25 + 4 = 29$.

The equation would be:

$$x^2 + (\text{opposite of the sum})x + \text{product} = 0$$
$$x^2 - 10x + 29 = 0$$

Find a quadratic equation with roots of $2/3$ and $^-3/4$.

MATHEMATICS HIGH SCHOOL

Solutions:

The sum of $2/3$ and $^-3/4$ is $^-1/12$. The product of $2/3$ and $^-3/4$ is $^-1/2$. The equation would be :

$$x^2 + (\text{opposite of the sum})x + \text{product} = 0$$
$$x^2 + 1/12\, x - 1/2 = 0$$

Common denominator = 12, so multiply by 12.

$$12(x^2 + 1/12\, x - 1/2 = 0$$
$$12x^2 + 1x - 6 = 0$$
$$12x^2 + x - 6 = 0$$

Try these:

Find a quadratic equation with a root of 5.
Find a quadratic equation with roots of $8/5$ and $^-6/5$.
Find a quadratic equation with roots of 12 and $^-3$.

SKILL 13.5 Identify quadratic equalities and inequalities for solving word problems.

Some word problems can be solved by setting up a quadratic equation or inequality. Examples of this type could be problems that deal with finding a maximum area. Examples follow:

Example 1:

A family wants to enclose 3 sides of a rectangular garden with 200 feet of fence. In order to have the maximum area possible, find the dimensions of the garden. Assume that the fourth side of the garden is already bordered by a wall or a fence.

Solution:
Let x = distance from the wall

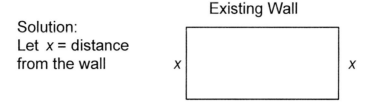

Existing Wall

Then $2x$ feet of fence is used for these 2 sides. The remaining side of the garden would use the rest of the 200 feet of fence, that is, $200-2x$ feet of fence. Therefore the width of the garden is x feet and the length is $200-2x$ ft. The area, called y, would equal length times width:

$$y = x(200-2x) = 200x - 2x^2$$

In this equation, a = $^-2$, b = 200, c = 0. The maximum area of this garden would occur at the vertex, where $x = {^-b}/{2a}$.
Substituting for a and b in this equation, this equation becomes
$x = {^-200}/(2 \times {^-2}) = {^-200}/({^-4}) = 50$ feet. If x = 50 ft, then $200-2x$ equals the other 100 feet. The maximum area occurs when the length is 100 feet and each of the widths is 50 feet. The maximum area = $100 \times 50 = 5000$ square feet.

Example 2:

A family wants to enclose 3 sides of a rectangular garden with 200 feet of fence. In order to have a garden with an area of **at least** 4800 square feet, find the dimensions of the garden. Assume that the fourth side of the garden is already bordered by a wall or a fence.

Solution:
Let x = distance from the wall

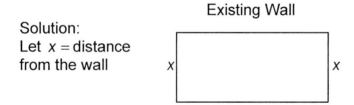

Existing Wall

Then 2x feet of fence is used for these 2 sides. The remaining side of the garden would use the rest of the 200 feet of fence, that is, $200-2x$ feet of fence. Therefore the width of the garden is x feet and the length is $200-2x$ ft. The area, called y, would have to be greater than or equal to the length times the width:

$$y \geq x(200-2x) \text{ or } y \geq 200x - 2x^2$$

In this equation, a = $^-2$, b = 200, c = 0. The area, $200x - 2x^2$, needs to be greater than or equal to 4800 sq. ft. So, this problem uses the inequality $4800 \leq 200x - 2x^2$. This becomes $2x^2 - 200x + 4800 \leq 0$. Solving this, we get:

$$2(x^2 - 100x + 2400) \leq 0$$
$$2(x - 60)(x - 40) \leq 0$$

If x = 60 or x = 40, then the area is at least 4800 sq. ft.
So the area will be at least 4800 square feet if the width of the garden is from 40 up to 60 feet. (The length of the rectangle would vary from 120 feet to 80 feet depending on the width of the garden.)

SKILL 13.6 Graph relations involving quadratics and estimate zeros from the graphs.

Quadratic equations can be used to model different real life situations. The graphs of these quadratics can be used to determine information about this real life situation.

Example:
The height of a projectile fired upward at a velocity of v meters per second from an original height of h meters is $y = h + vx - 4.9x^2$. If a rocket is fired from an original height of 250 meters with an original velocity of 4800 meters per second, find the approximate time the rocket would drop to sea level (a height of 0).
Solution:
The equation for this problem is: $y = 250 + 4800x - 4.9x^2$. If the height at sea level is zero, then y = 0 so $0 = 250 + 4800x - 4.9x^2$. Solving this for x could be done by using the quadratic formula. In addition, the approximate time in x seconds until the rocket would be at sea level could be estimated by looking at the graph. When the y value of the graph goes from positive to negative then there is a root (also called solution or x intercept) in that interval.

$$x = \frac{^-4800 \pm \sqrt{4800^2 - 4(^-4.9)(250)}}{2(^-4.9)} \approx 980 \text{ or } ^-0.05 \text{ seconds}$$

Since the time has to be positive, it will be about 980 seconds until the rocket is at sea level.

SKILL 13.7 Graph relations involving quadratic inequalities.

To graph an inequality, graph the quadratic as if it was an equation; however, if the inequality has just a > or < sign, then make the curve itself dotted. Shade above the curve for > or ≥. Shade below the curve for < or ≤.

Examples:

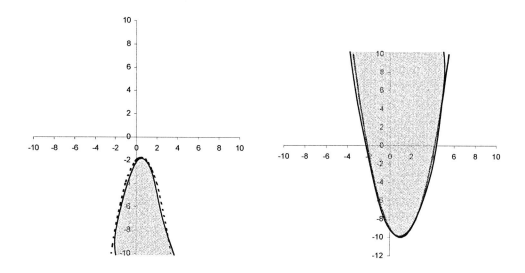

TEACHER CERTIFICATION EXAM

COMPETENCY 14.0   **PROFICIENCY IN EVALUATING AND GRAPHING FUNCTIONS.**

SKILL 14.1   **Graph simple polynomial functions and estimate zeros from the graphs.**

There are 2 easy ways to find the values of a function. First to find the value of a function when $x = 3$, substitute 3 in place of every letter x. Then simplify the expression following the order of operations. For example, if $f(x) = x^3 - 6x + 4$, then to find f(3), substitute 3 for x.
The equation becomes $f(3) = 3^3 - 6(3) + 4 = 27 - 18 + 4 = 13$.
So (3, 13) is a point on the graph of f(x).

A second way to find the value of a function is to use synthetic division. To find the value of a function when $x = 3$, divide 3 into the coefficients of the function. (Remember that coefficients of missing terms, like x^2, must be included). The remainder is the value of the function.
If $f(x) = x^3 - 6x + 4$, then to find f(3) using synthetic division:

Note the 0 for the missing x^2 term.

$$\begin{array}{r|rrrr} 3 & 1 & 0 & ^-6 & 4 \\ & & 3 & 9 & 9 \\ \hline & 1 & 3 & 3 & 13 \end{array}$$ ← this is the value of the function.

Therefore, (3, 13) is a point on the graph of $f(x) = x^3 - 6x + 4$.

Example: Find values of the function at integer values from x = -3 to x = 3 if $f(x) = x^3 - 6x + 4$.

If $x = {}^-3$:

$$f(^-3) = (^-3)^3 - 6(^-3) + 4$$
$$= (^-27) - 6(^-3) + 4$$
$$= {}^-27 + 18 + 4 = {}^-5$$

MATHEMATICS HIGH SCHOOL

synthetic division:

$$\begin{array}{r|rrrr} -3 & 1 & 0 & -6 & 4 \\ & & -3 & 9 & -9 \\ \hline & 1 & -3 & 3 & -5 \end{array}$$ ← this is the value of the function if $x = {}^-3$.
Therefore, $({}^-3, {}^-5)$ is a point on the graph.

If $x = {}^-2$:

$$f({}^-2) = ({}^-2)^3 - 6({}^-2) + 4$$
$$= ({}^-8) - 6({}^-2) + 4$$
$$= {}^-8 + 12 + 4 = 8 \leftarrow \text{this is the value of the function if } x = {}^-2.$$
Therefore, $({}^-2, 8)$ is a point on the graph.

If $x = {}^-1$:

$$f({}^-1) = ({}^-1)^3 - 6({}^-1) + 4$$
$$= ({}^-1) - 6({}^-1) + 4$$
$$= {}^-1 + 6 + 4 = 9$$

synthetic division:

$$\begin{array}{r|rrrr} -1 & 1 & 0 & -6 & 4 \\ & & -1 & 1 & 5 \\ \hline & 1 & -1 & -5 & 9 \end{array}$$ ← this is the value if the function if $x = {}^-1$.
Therefore, $({}^-1, 9)$ is a point on the graph.

If $x = 0$:

$$f(0) = (0)^3 - 6(0) + 4$$
$$= 0 - 6(0) + 4$$
$$= 0 - 0 + 4 = 4 \leftarrow \text{this is the value of the function if } x = 0.$$
Therefore, $(0, 4)$ is a point on the graph.

If $x = 1$:

$$f(1) = (1)^3 - 6(1) + 4$$
$$= (1) - 6(1) + 4$$
$$= 1 - 6 + 4 = {}^-1$$

synthetic division:

$$\begin{array}{r|rrrr} 1 & 1 & 0 & {}^-6 & 4 \\ & & 1 & 1 & {}^-5 \\ \hline & 1 & 1 & {}^-5 & {}^-1 \end{array}$$

1   1   ⁻5   ⁻1 ← this is the value if the function of $x = 1$.
Therefore, $(1, {}^-1)$ is a point on the graph.

If $x = 2$:

$$f(2) = (2)^3 - 6(2) + 4$$
$$= 8 - 6(2) + 4$$
$$= 8 - 12 + 4 = 0$$

synthetic division:

$$\begin{array}{r|rrrr} 2 & 1 & 0 & {}^-6 & 4 \\ & & 2 & 4 & {}^-4 \\ \hline & 1 & 2 & {}^-2 & 0 \end{array}$$

1   2   ⁻2   0 ← this is the value of the function if $x = 2$.
Therefore, $(2, 0)$ is a point on the graph.

If $x = 3$:

$$f(3) = (3)^3 - 6(3) + 4$$
$$= 27 - 6(3) + 4$$
$$= 27 - 18 + 4 = 13$$

synthetic division:

$$\begin{array}{r|rrrr} 3 & 1 & 0 & {}^-6 & 4 \\ & & 3 & 9 & 9 \\ \hline & 1 & 3 & 3 & 13 \end{array}$$

1   3   3   13 ← this is the value of the function if $x = 3$.
Therefore, $(3, 13)$ is a point on the graph.

The following points are points on the graph:

X	Y
−3	−5
−2	8
−1	9
0	4
1	−1
2	0
3	13

Note the change in sign of the y value between $x = {}^-3$ and $x = {}^-2$. This indicates there is a zero between $x = {}^-3$ and $x = {}^-2$. Since there is another change in sign of the y value between $x = 0$ and $x = {}^-1$, there is a second root there. When $x = 2$, $y = 0$ so $x = 2$ is an exact root of this polynomial.

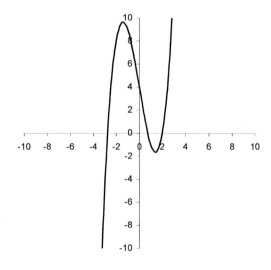

TEACHER CERTIFICATION EXAM

SKILL 14.2 **Find the values of a given polynomial function over a specified domain.**

Explanation is found in SKILL 14.1.

Example: Find values of the function at $x = ^-5, 2,$ and 17 if
$f(x) = 2x^5 - 4x^3 + 3x^2 - 9x + 10$.

If $x = ^-5$:

$$f(^-5) = 2(^-5)^5 - 4(^-5)^3 + 3(^-5)^2 - 9(^-5) + 10$$
$$= 2(^-3125) - 4(^-125) + 3(25) - 9(^-5) + 10$$
$$= ^-6250 + 500 + 75 + 45 + 10 = ^-5620$$

synthetic division:

$$\begin{array}{r|rrrrrr} ^-5 & 2 & 0 & ^-4 & 3 & ^-9 & 10 \\ & & ^-10 & 50 & ^-230 & 1135 & ^-5630 \\ \hline & 2 & ^-10 & 46 & ^-227 & ^-1126 & ^-5620 \end{array}$$

← this is the value of the function if $x = ^-5$.
Therefore, $(^-5, ^-5620)$ is a point on the graph.

If $x = 2$:

$$f(2) = 2(2)^5 - 4(2)^3 + 3(2)^2 - 9(2) + 10$$
$$= 2(32) - 4(8) + 3(4) - 9(2) + 10$$
$$= 64 - 32 + 12 - 18 + 10 = 36$$

synthetic division:

$$\begin{array}{r|rrrrrr} 2 & 2 & 0 & ^-4 & 3 & ^-9 & 10 \\ & & 4 & 8 & 8 & 22 & 26 \\ \hline & 2 & 4 & 4 & 11 & 13 & 36 \end{array}$$

← this is the value of the function if $x = 2$.
Therefore, $(2, 36)$ is a point on the graph.

MATHEMATICS HIGH SCHOOL

If $x = 17$:

$$f(17) = 2(17)^5 - 4(17)^3 + 3(17)^2 - 9(17) + 10$$
$$= 2(1419857) - 4(4913) + 3(289) - 9(17) + 10$$
$$= 2839714 - 19652 + 867 - 153 + 10 = 2820786$$

synthetic division:

$$17 \underline{| \begin{array}{cccccc} 2 & 0 & ^-4 & 3 & ^-9 & 10 \\ & 34 & 578 & 9758 & 165937 & 2820776 \end{array}}$$
$$\; 2 \;\; 34 \;\; 574 \;\; 9761 \;\; 165928 \;\; 2820786 \leftarrow \text{this is the value}$$

of the function if $x = 17$.
Therefore, (17, 2820786) is a point on the graph.

SKILL 14.3 Apply the first-degree equations in word problems with two variables.

Word problems can sometimes be solved by using a system of two equations in 2 unknowns. This system can then be solved using substitution, the addition-subtraction method, or graphing.

Example: Mrs. Winters bought 4 dresses and 6 pairs of shoes for $340. Mrs. Summers went to the same store and bought 3 dresses and 8 pairs of shoes for $360. If all the dresses were the same price and all the shoes were the same price, find the price charged for a dress and for a pair of shoes.

Let x = price of a dress
Let y = price of a pair of shoes

Then Mrs. Winters' equation would be: $4x + 6y = 340$
Mrs. Summers' equation would be: $3x + 8y = 360$

TEACHER CERTIFICATION EXAM

To solve by addition-subtraction:

Multiply the first equation by 4: $4(4x + 6y = 340)$
Multiply the other equation by $^-3$: $^-3(3x + 8y = 360)$
By doing this, the equations can be added to each other to eliminate one variable and solve for the other variable.

$$16x + 24y = 1360$$
$$\underline{-9x - 24y = {}^-1080}$$
$$7x = 280$$
$$x = 40 \leftarrow \text{the price of a dress was \$40}$$

solving for y, $y = 30$ ← the price of a pair of shoes, $30

Example: Aardvark Taxi charges $4 initially plus $1 for every mile traveled. Baboon Taxi charges $6 initially plus $.75 for every mile traveled. Determine when it is cheaper to ride with Aardvark Taxi or to ride with Baboon Taxi.

Aardvark Taxi's equation: $y = 1x + 4$
Baboon Taxi's equation : $y = .75x + 6$

Using substitution: $.75x + 6 = x + 4$
Multiplying by 4: $3x + 24 = 4x + 16$
Solving for x : $8 = x$

This tells you that at 8 miles the total charge for the 2 companies is the same. If you compare the charge for 1 mile, Aardvark charges $5 and Baboon charges $6.75. Clearly Aardvark is cheaper for distances up to 8 miles, but Baboon Taxi is cheaper for distances greater than 8 miles.

This problem can also be solved by graphing the 2 equations.

$y = 1x + 4$ $y = .75x + 6$

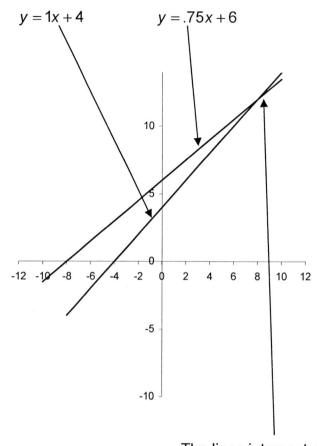

The lines intersect at (8, 12), therefore at 8 miles both companies charge $12. At values less than 8 miles, Aardvark Taxi charges less (the graph is below Baboon). Greater than 8 miles, Aardvark charges more (the graph is above Baboon).

TEACHER CERTIFICATION EXAM

COMPETENCY 15.0 COMPREHENSION OF DIRECT AND INVERSE VARIATIONS.

SKILL 15.1 Solve problems using direct and inverse variations.

-If two things vary directly, as one gets larger, the other also gets larger. If one gets smaller, then the other gets smaller too. If x and y vary directly, there should be a constant, c, such that $y = cx$. Something can also vary directly with the square of something else, $y = cx^2$.

-If two things vary inversely, as one gets larger, the other one gets smaller instead. If x and y vary inversely, there should be a constant, c, such that $xy = c$ or $y = c/x$. Something can also vary inversely with the square of something else, $y = c/x^2$.

Example: If $30 is paid for 5 hours work, how much would be paid for 19 hours work?

This is direct variation and $30 = 5c, so the constant is 6 ($6/hour). So $y = 6(19)$ or $y = \$114$.
This could also be done
as a proportion: $\dfrac{\$30}{5} = \dfrac{y}{19}$

$5y = 570$
$y = 114$

Example: On a 546 mile trip from Miami to Charlotte, one car drove 65 mph while another car drove 70 mph. How does this affect the driving time for the trip?

This is an inverse variation, since increasing your speed should decrease your driving time. Using the equation: rate × time = distance, rt = d.

65t = 546   and   70t = 546
t = 8.4   and   t = 7.8
slower speed, more time   faster speed, less time

MATHEMATICS HIGH SCHOOL

Example: A 14" pizza from Azzip Pizza costs $8.00. How much would a 20" pizza cost if its price was based on the same price per square inch?

Here the price is directly proportional to the square of the radius. Using a proportion:

$$\frac{\$8.00}{7^2 \pi} = \frac{x}{10^2 \pi}$$

$$\frac{8}{153.86} = \frac{x}{314}$$

$$16.33 = x$$

$16.33 would be the price of the large pizza.

TEACHER CERTIFICATION EXAM

COMPETENCY 16.0 ABILITY TO SOLVE EQUATIONS AND INEQUALITIES

SKILL 16.1 Solve systems of linear equations with two or three variables.

Word problems can sometimes be solved by using a system of two equations in 2 unknowns. This system can then be solved using **substitution**, the **addition-subtraction method**, or **determinants**.

Example: Farmer Greenjeans bought 4 cows and 6 sheep for $1700. Mr. Ziffel bought 3 cows and 12 sheep for $2400. If all the cows were the same price and all the sheep were another price, find the price charged for a cow or for a sheep.

Let x = price of a cow
Let y = price of a sheep

Then Farmer Greenjeans' equation would be: $4x + 6y = 1700$
Mr. Ziffel's equation would be: $3x + 12y = 2400$

To solve by **addition-subtraction**:
Multiply the first equation by $^-2$: $^-2(4x + 6y = 1700)$
Keep the other equation the same : $(3x + 12y = 2400)$
By doing this, the equations can be added to each other to eliminate one variable and solve for the other variable.

$$^-8x - 12y = ^-3400$$
$$\underline{3x + 12y = 2400} \quad \text{Add these equations.}$$
$$^-5x \quad\quad = ^-1000$$

$x = 200 \leftarrow$ the price of a cow was $200.
Solving for y, $y = 150 \leftarrow$ the price of a sheep, $150.

To solve by **substitution**:
Solve one of the equations for a variable. (Try to make an equation without fractions if possible.) Substitute this expression into the equation that you have not yet used. Solve the resulting equation for the value of the remaining variable.

$$4x + 6y = 1700$$
$$3x + 12y = 2400 \leftarrow \text{Solve this equation for } x.$$

It becomes $x = 800 - 4y$. Now substitute $800 - 4y$ in place of x in the OTHER equation. $4x + 6y = 1700$ now becomes:

$$4(800 - 4y) + 6y = 1700$$
$$3200 - 16y + 6y = 1700$$
$$3200 - 10y = 1700$$
$$^-10y = ^-1500$$
$$y = 150, \text{ or } \$150 \text{ for a sheep.}$$

Substituting 150 back into an equation for y, find x.
$$4x + 6(150) = 1700$$
$$4x + 900 = 1700$$
$$4x = 800 \text{ so } x = 200 \text{ for a cow.}$$

To solve by **determinants**:

Let x = price of a cow
Let y = price of a sheep

Then Farmer Greenjeans' equation would be: $4x + 6y = 1700$
Mr. Ziffel's equation would be: $3x + 12y = 2400$

To solve this system using determinants, make one 2 by 2 determinant divided by another 2 by 2 determinant. The bottom determinant is filled with the x and y term coefficients. The top determinant is almost the same as this bottom determinant. The only difference is that when you are solving for x, the x coefficients are replaced with the constants. Likewise, when you are solving for y, the y coefficients are replaced with the constants. To find the value of a 2 by 2 determinant, $\begin{pmatrix} a & b \\ c & d \end{pmatrix}$, is found by $ad - bc$.

$$x = \frac{\begin{pmatrix} 1700 & 6 \\ 2400 & 12 \end{pmatrix}}{\begin{pmatrix} 4 & 6 \\ 3 & 12 \end{pmatrix}} = \frac{1700(12) - 6(2400)}{4(12) - 6(3)} = \frac{20400 - 14400}{48 - 18} = \frac{6000}{30} = 200$$

$$y = \frac{\begin{pmatrix} 4 & 1700 \\ 3 & 2400 \end{pmatrix}}{\begin{pmatrix} 4 & 6 \\ 3 & 12 \end{pmatrix}} = \frac{2400(4) - 3(1700)}{4(12) - 6(3)} = \frac{9600 - 5100}{48 - 18} = \frac{4500}{30} = 150$$

NOTE: The bottom determinant is always the same value for each letter.

Word problems can sometimes be solved by using a system of three equations in 3 unknowns. This system can then be solved using **substitution**, the **addition-subtraction method**, or **determinants**.

To solve by **substitution**:

Example: Mrs. Allison bought 1 pound of potato chips, a 2 pound beef roast, and 3 pounds of apples for a total of $ 8.19. Mr. Bromberg bought a 3 pound beef roast and 2 pounds of apples for $ 9.05. Kathleen Kaufman bought 2 pounds of potato chips, a 3 pound beef roast, and 5 pounds of apples for $ 13.25. Find the per pound price of each item.

Let x = price of a pound of potato chips
Let y = price of a pound of roast beef
Let z = price of a pound of apples

Mrs. Allison's equation would be: $1x + 2y + 3z = 8.19$
Mr. Bromberg's equation would be: $3y + 2z = 9.05$
K. Kaufman's equation would be: $2x + 3y + 5z = 13.25$

Take the first equation and solve it for x. (This was chosen because x is the easiest variable to get alone in this set of equations.) This equation would become:

$$x = 8.19 - 2y - 3z$$

Substitute this expression into the other equations in place of the letter x:

$$3y + 2z = 9.05 \leftarrow \text{equation 2}$$
$$2(8.19 - 2y - 3z) + 3y + 5z = 13.25 \leftarrow \text{equation 3}$$

TEACHER CERTIFICATION EXAM

Simplify the equation by combining like terms:

$$3y + 2z = 9.05 \leftarrow \text{equation 2}$$
$$*\ {}^-1y - 1z = {}^-3.13 \leftarrow \text{equation 3}$$

Solve equation 3 for either y or z:

$y = 3.13 - z$ Substitute this into equation 2 for y:

$$3(3.13 - z) + 2z = 9.05 \leftarrow \text{equation 2}$$
$${}^-1y - 1z = {}^-3.13 \leftarrow \text{equation 3}$$

Combine like terms in equation 2:

$$9.39 - 3z + 2z = 9.05$$
$$z = .34 \quad \text{per pound price of bananas}$$

Substitute .34 for z in the starred equation above to solve for y:

$$y = 3.13 - z \text{ becomes } y = 3.13 - .34, \text{ so}$$
$$y = 2.79 = \text{per pound price of roast beef}$$

Substituting .34 for z and 2.79 for y in one of the original equations, solve for x:

$$1x + 2y + 3z = 8.19$$
$$1x + 2(2.79) + 3(.34) = 8.19$$
$$x + 5.58 + 1.02 = 8.19$$
$$x + 6.60 = 8.19$$
$$x = 1.59 \text{ per pound of potato chips}$$

$(x, y, z) = (1.59, 2.79, .34)$

MATHEMATICS HIGH SCHOOL

TEACHER CERTIFICATION EXAM

To solve by **addition-subtraction**:

Choose a letter to eliminate. Since the second equation is already missing an x, let's eliminate x from equations 1 and 3.

1) $1x + 2y + 3x = 8.19$ ← Multiply by $^-2$ below.
2) $\quad\quad 3y + 2z = 9.05$
3) $2x + 3y + 5z = 13.25$

$^-2(1x + 2y + 3z = 8.19) \quad = \quad ^-2x - 4y - 6z = {}^-16.38$
Keep equation 3 the same : $\quad\quad 2x + 3y + 5z = 13.25$

By doing this, the equations $\quad\quad\quad {}^-y - z = {}^-3.13$ ← equation 4
can be added to each other to
eliminate one variable.

The equations left to solve are equations 2 and 4:
$\quad {}^-y - z = {}^-3.13$ ← equation 4
$\quad 3y + 2z = 9.05$ ← equation 2

Multiply equation 4 by 3: $\quad 3({}^-y - z = {}^-3.13)$
Keep equation 2 the same: $\quad 3y + 2z = 9.05$

$\quad\quad {}^-3y - 3z = {}^-9.39$
$\quad\quad 3y + 2z = 9.05 \quad\quad$ Add these equations.
$\quad\quad\quad {}^-1z = {}^-.34$
$\quad\quad\quad\quad z = .34$ ← the per pound price of bananas
solving for y, $y = 2.79$ ← the per pound roast beef price
solving for x, $x = 1.59$ ← potato chips, per pound price

To solve by **substitution**:

Solve one of the 3 equations for a variable. (Try to make an equation without fractions if possible.) Substitute this expression into the other 2 equations that you have not yet used.

1) $1x + 2y + 3z = 8.19$ ← Solve for x.
2) $\quad\quad 3y + 2z = 9.05$
3) $2x + 3y + 5z = 13.25$
Equation 1 becomes $x = 8.19 - 2y - 3z$.

MATHEMATICS HIGH SCHOOL

Substituting this into equations 2 and 3, they become:

2) $\quad 3y + 2z = 9.05$
3) $\quad 2(8.19 - 2y - 3z) + 3y + 5z = 13.25$
$\quad\quad 16.38 - 4y - 6z + 3y + 5z = 13.25$
$\quad\quad {}^-y - z = {}^-3.13$

The equations left to solve are:

$\quad 3y + 2z = 9.05$
$\quad {}^-y - z = {}^-3.13 \leftarrow$ Solve for either y or z.

It becomes $y = 3.13 - z$. Now substitute $3.13 - z$ in place of y in the OTHER equation. $3y + 2z = 9.05$ now becomes:

$$3(3.13 - z) + 2z = 9.05$$
$$9.39 - 3z + 2z = 9.05$$
$$9.39 - z = 9.05$$
$${}^-z = {}^-.34$$
$$z = .34, \text{ or } \$.34/\text{lb of bananas}$$

Substituting .34 back into an equation for z, find y.
$\quad 3y + 2z = 9.05$
$\quad 3y + 2(.34) = 9.05$
$\quad 3y + .68 = 9.05$ so $y = 2.79/\text{lb of roast beef}$

Substituting .34 for z and 2.79 for y into one of the original equations, it becomes:

$\quad 2x + 3y + 5z = 13.25$
$\quad 2x + 3(2.79) + 5(.34) = 13.25$
$\quad 2x + 8.37 + 1.70 = 13.25$
$\quad 2x + 10.07 = 13.25$, so $x = 1.59/\text{lb of potato chips}$

To solve by **determinants**:

Let x = price of a pound of potato chips
Let y = price of a pound of roast beef
Let z = price of a pound of apples

1) $1x + 2y + 3z = 8.19$
2) $3y + 2z = 9.05$
3) $2x + 3y + 5z = 13.25$

To solve this system using determinants, make one 3 by 3 determinant divided by another 3 by 3 determinant. The bottom determinant is filled with the x, y, and z term coefficients. The top determinant is almost the same as this bottom determinant. The only difference is that when you are solving for x, the x coefficients are replaced with the constants. When you are solving for y, the y coefficients are replaced with the constants. Likewise, when you are solving for z, the z coefficients are replaced with the constants. To find the value of a 3 by 3 determinant,

$$\begin{pmatrix} a & b & c \\ d & e & f \\ g & h & i \end{pmatrix}$$ is found by the following steps:

Copy the first two columns to the right of the determinant:

$$\begin{pmatrix} a & b & c \\ d & e & f \\ g & h & i \end{pmatrix} \begin{matrix} a & b \\ d & e \\ g & h \end{matrix}$$

Multiply the diagonals from top left to bottom right, and add these diagonals together.

$$\begin{pmatrix} a^* & b^\circ & c^\bullet \\ d & e^* & f^\circ \\ g & h & i^* \end{pmatrix} \begin{matrix} a & b \\ d^\bullet & e \\ g^\circ & h^\bullet \end{matrix} = a^*e^*i^* + b^\circ f^\circ g^\circ + c^\bullet d^\bullet h^\bullet$$

Then multiply the diagonals from bottom left to top right, and add these diagonals together.

TEACHER CERTIFICATION EXAM

$$\begin{pmatrix} a & b & c^* \\ d & e^* & f^\circ \\ g^* & h^\circ & i^\bullet \end{pmatrix} \begin{matrix} a^\circ & b^\bullet \\ d^\bullet & e \\ g & h \end{matrix} = g^*e^*c^* + h^\circ f^\circ a^\circ + i^\bullet d^\bullet b^\bullet$$

Subtract the first diagonal total minus the second diagonal total:

$$(= a^*e^*i^* + b^\circ f^\circ g^\circ + c^\bullet d^\bullet h^\bullet) - (= g^*e^*c^* + h^\circ f^\circ a^\circ + i^\bullet d^\bullet b^\bullet)$$

This gives the value of the determinant. To find the value of a variable, divide the value of the top determinant by the value of the bottom determinant.

1) $1x + 2y + 3z = 8.19$
2) $3y + 2z = 9.05$
3) $2x + 3y + 5z = 13.25$

$$x = \frac{\begin{pmatrix} 8.19 & 2 & 3 \\ 9.05 & 3 & 2 \\ 13.25 & 3 & 5 \end{pmatrix}}{\begin{pmatrix} 1 & 2 & 3 \\ 0 & 3 & 2 \\ 2 & 3 & 5 \end{pmatrix}}$$ solve each determinant using the method shown below

Multiply the diagonals from top left to bottom right, and add these diagonals together.

$$\begin{pmatrix} 8.19^* & 2^\circ & 3^\bullet \\ 9.05 & 3^* & 2^\circ \\ 13.25 & 3 & 5^* \end{pmatrix} \begin{matrix} 8.19 & 2 \\ 9.05^\bullet & 3 \\ 13.25^\circ & 3^\bullet \end{matrix} = (8.19^*)(3^*)(5^*) + (2^\circ)(2^\circ)(13.25^\circ) + (3^\bullet)(9.05^\bullet)(3^\bullet)$$

Then multiply the diagonals from bottom left to top right, and add these diagonals together.

$$\begin{pmatrix} 8.19 & 2 & 3^* \\ 9.05 & 3^* & 2^\circ \\ 13.25^* & 3^\circ & 5^\bullet \end{pmatrix} \begin{matrix} 8.19^\circ & 2^\bullet \\ 9.05^\bullet & 3 \\ 13.25 & 3 \end{matrix}$$

$$= (13.25^*)(3^*)(3^*) + (3^\circ)(2^\circ)(8.19^\circ) + (5^\bullet)(9.05^\bullet)(2^\bullet)$$

Subtract the first diagonal total minus the second diagonal total:

$(8.19^*)(3^*)(5^*) + (2^\circ)(2^\circ)(13.25^\circ) + (3^\bullet)(9.05^\bullet)(3^\bullet) = 257.30$

$- (13.25^*)(3^*)(3^*) + (3^\circ)(2^\circ)(8.19^\circ) + (5^\bullet)(9.05^\bullet)(2^\bullet) = {}^-258.89$

${}^-1.59$

Do same multiplying and subtraction procedure for the bottom determinant to get $^-1$ as an answer. Now divide:

$$\frac{^-1.59}{^-1} = \$1.59/\text{lb of potato chips}$$

$$y = \frac{\begin{pmatrix} 1 & 8.19 & 3 \\ 0 & 9.05 & 2 \\ 2 & 13.25 & 5 \end{pmatrix}}{\begin{pmatrix} 1 & 2 & 3 \\ 0 & 3 & 2 \\ 2 & 3 & 5 \end{pmatrix}} = \frac{^-2.79}{^-1} = \$2.79/\text{lb of roast beef}$$

NOTE: The bottom determinant is always the same value for each letter.

$$z = \frac{\begin{pmatrix} 1 & 2 & 8.19 \\ 0 & 3 & 9.05 \\ 2 & 3 & 13.25 \end{pmatrix}}{\begin{pmatrix} 1 & 2 & 3 \\ 0 & 3 & 2 \\ 2 & 3 & 5 \end{pmatrix}} = \frac{^-.34}{^-1} = \$.34/\text{lb of bananas}$$

TEACHER CERTIFICATION EXAM

SKILL 16.2 Solve systems of linear inequalities graphically.

To graph an inequality, solve the inequality for y. This gets the inequality in **slope intercept form**, (for example : $y < mx + b$). The point (0,b) is the y-intercept and m is the line's slope.

- If the inequality solves to $x \geq$ **any number**, then the graph includes a **vertical line**.

- If the inequality solves to $y \leq$ **any number**, then the graph includes a **horizontal line**.

- When graphing a linear inequality, the line will be dotted if the inequality sign is < or >. If the inequality signs are either $\geq$ or $\leq$, the line on the graph will be a solid line. Shade above the line when the inequality sign is $\geq$ or >. Shade below the line when the inequality sign is < or $\leq$. For inequalities of the forms $x >$ number, $x \leq$ number, $x <$ number, or $x \geq$ number, draw a vertical line (solid or dotted). Shade to the right for > or $\geq$. Shade to the left for < or $\leq$.

> Remember: **Dividing or multiplying by a negative number will reverse the direction of the inequality sign.**

Use these rules to graph and shade each inequality. The solution to a system of linear inequalities consists of the part of the graph that is shaded for each inequality. For instance, if the graph of one inequality was shaded with red, and the graph of another inequality was shaded with blue, then the overlapping area would be shaded purple. The purple area would be the points in the solution set of this system.

Example: Solve by graphing:

$x + y \leq 6$
$x - 2y \leq 6$

Solving the inequalities for y, they become:

$y \leq {}^-x + 6$ (y intercept of 6 and slope = $^-1$)
$y \geq 1/2\, x - 3$ (y intercept of $^-3$ and slope = 1/2)

MATHEMATICS HIGH SCHOOL

A graph with shading is shown below:

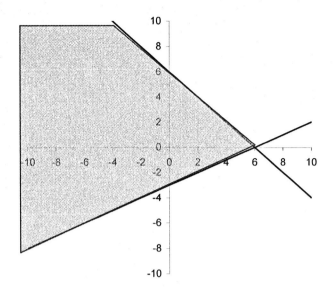

SKILL 16.3 Solve linear equalities or inequalities with one variable over rational numbers.

To solve an equation or inequality, follow these steps:

STEP 1. If there are parentheses, use the distributive property to eliminate them.

STEP 2. If there are fractions, determine their LCD (least common denominator). Multiply every term of the equation by the LCD. This will cancel out all of the fractions while solving the equation or inequality.

STEP 3. If there are decimals, find the longest decimal. Multiply each term by a power of 10(10, 100, 1000,etc.) with the same number of zeros as the length of the decimal. This will eliminate all decimals while solving the equation or inequality.

STEP 4. Combine like terms on each side of the equation or inequality.

STEP 5. If there are variables on both sides of the equation, add or subtract one of those variable terms to move it to the other side. Combine like terms.

STEP 6. If there are constants on both sides, add or subtract one of those constants to move it to the other side. Combine like terms.

TEACHER CERTIFICATION EXAM

STEP 7. If there is a coefficient in front of the variable, divide both sides by this number. This is the answer to an equation. However, remember:

Dividing or multiplying an inequality by a negative number will reverse the direction of the inequality sign.

STEP 8. The solution of a linear equation solves to one single number. The solution of an inequality is always stated including the inequality sign.

Example: Solve: $3(2x+5)-4x=5(x+9)$

$6x+15-4x=5x+45$	ref. step 1
$2x+15=5x+45$	ref. step 4
$^{-}3x+15=45$	ref. step 5
$^{-}3x=30$	ref. step 6
$x=\ ^{-}10$	ref. step 7

Example: Solve: $1/2(5x+34)=1/4(3x-5)$

$5/2\,x+17=3/4\,x-5/4$	ref. step 1
LCD of 5/2, 3/4, and 5/4 is 4.	
Multiply by the LCD of 4.	
$4(5/2\,x+17=3/4\,x-5/4)$	ref. step 2
$10x+68=3x-5$	
$7x+68=\ ^{-}5$	ref. step 5
$7x=\ ^{-}73$	ref. step 6
$x=\ ^{-}73/10$ or $^{-}7\ 3/10$	ref. step 7

Example: Solve: $6x+21<8x+31$

$^{-}2x+21<31$	ref. step 5
$^{-}2x<10$	ref. step 6
$x>\ ^{-}5$	ref. step 7

Note that the inequality sign has changed.

MATHEMATICS HIGH SCHOOL

SKILL 16.4 Solve linear equalities or inequalities involving absolute values with one variable.

To solve an absolute value equation, follow these steps:

1. Get the absolute value expression alone on one side of the equation.

2. Split the absolute value equation into 2 separate equations without absolute value bars. Write the expression inside the absolute value bars (without the bars) equal to the expression on the other side of the equation. Now write the expression inside the absolute value bars equal to the opposite of the expression on the other side of the equation.

3. Now solve each of these equations.

4. **Check each answer by substituting them into the original equation** (with the absolute value symbol). There will be answers that do not check in the original equation. These answers are discarded as they are **extraneous solutions**. If all answers are discarded as incorrect, then the answer to the equation is $\varnothing$, which means the empty set or the null set. (0, 1, or 2 solutions could be correct.)

To solve an absolute value inequality, follow these steps:

1. Get the absolute value expression alone on one side of the inequality. Remember: **Dividing or multiplying by a negative number will reverse the direction of the inequality sign.**

2. Remember what the inequality sign is at this point.

3. Split the absolute value inequality into 2 separate inequalities without absolute value bars. First rewrite the inequality without the absolute bars and solve it. Next write the expression inside the absolute value bar followed by the opposite inequality sign and then by the opposite of the expression on the other side of the inequality. Now solve it.

4. If the sign in the inequality on step 2 is $<$ or $\leq$, the answer is those 2 inequalities connected by the word **and**. The solution set consists of the points between the 2 numbers on the number line. If the sign in the inequality on step 2 is $>$ or $\geq$, the answer is those 2 inequalities connected by the word **or**. The solution set consists of the points outside the 2 numbers on the number line.

If an expression inside an absolute value bar is compared to a negative number, the answer can also be either all real numbers or the empty set ($\varnothing$). For instance,

$$|x+3| < {}^-6$$

would have the empty set as the answer, since an absolute value is always positive and will never be less than $^-6$. However,

$$|x+3| > {}^-6$$

would have all real numbers as the answer, since an absolute value is always positive or at least zero, and will never be less than -6. In similar fashion,

$$|x+3| = {}^-6$$

would never check because an absolute value will never give a negative value.

<u>Example</u>: Solve and check:

$$|2x-5| + 1 = 12$$
$$|2x-5| = 11 \quad \text{Get absolute value alone.}$$

Rewrite as 2 equations and solve separately.

same equation without absolute value		same equation without absolute value but right side is opposite
$2x - 5 = 11$		$2x - 5 = {}^-11$
$2x = 16$	and	$2x = {}^-6$
$x = 8$		$x = {}^-3$

Checks:
$$|2x-5| + 1 = 12 \qquad\qquad |2x-5| + 1 = 12$$
$$|2(8)-5| + 1 = 12 \qquad\qquad |2(^-3)-5| + 1 = 12$$
$$|11| + 1 = 12 \qquad\qquad |^-11| + 1 = 12$$
$$12 = 12 \qquad\qquad 12 = 12$$

This time both 8 and $^-3$ check.

Example: Solve and check:

$$2|x-7|-13 \geq 11$$

$$2|x-7| \geq 24 \quad \text{Get absolute value alone.}$$

$$|x-7| \geq 12$$

Rewrite as 2 inequalities and solve separately.

same inequality without absolute value		same inequality without absolute value but right side and inequality sign are both the opposite
$x - 7 \geq 12$	or	$x - 7 \leq {}^-12$
$x \geq 19$	or	$x \leq {}^-5$

SKILL 16.5 Identify equalities or inequalities that may be used to solve word problems involving one variable.

Equations and inequalities can be used to solve various types of word problems. Examples follow.

Example: The YMCA wants to sell raffle tickets to raise at least $32,000. If they must pay $7,250 in expenses and prizes out of the money collected from the tickets, how many tickets worth $25 each must they sell?

Solution: Since they want to raise **at least $32,000**, that means they would be happy to get 32,000 **or more**. This requires an inequality.

Let x = number of tickets sold
Then $25x$ = total money collected for x tickets

Total money minus expenses is greater than $32,000.

$$25x - 7250 \geq 32000$$
$$25x \geq 39250$$
$$x \geq 1570$$

TEACHER CERTIFICATION EXAM

If they sell **1,570 tickets or more**, they will raise AT LEAST $32,000.

Example: The Simpsons went out for dinner. All 4 of them ordered the aardvark steak dinner. Bert paid for the 4 meals and included a tip of $12 for a total of $84.60. How much was an aardvark steak dinner?

Let x = the price of one aardvark dinner.
So $4x$ = the price of 4 aardvark dinners.

$$4x + 12 = 84.60$$
$$4x = 72.60$$
$$x = \$18.50 \text{ for each dinner.}$$

SKILL 16.6 Identify equalities or inequalities that may be used to solve word problems involving one variable.

Some word problems can be solved using a system of equations or inequalities. Watch for words like greater than, less than, at least, or no more than which indicate the need for inequalities.

Example: Farmer Greenjeans bought 4 cows and 6 sheep for $1700. Mr. Ziffel bought 3 cows and 12 sheep for $2400. If all the cows were the same price and all the sheep were another price, find the price charged for a cow or for a sheep.

Let x = price of a cow
Let y = price of a sheep

Then Farmer Greenjeans' equation would be: $4x + 6y = 1700$
Mr. Ziffel's equation would be: $3x + 12y = 2400$

To solve by **addition-subtraction**:

Multiply the first equation by $^-2$: $^-2(4x+6y=1700)$
Keep the other equation the same : $(3x+12y=2400)$
By doing this, the equations can be added to each other to eliminate one variable and solve for the other variable.

$$^-8x-12y = {}^-3400$$
$$\underline{3x+12y = 2400} \qquad \text{Add these equations.}$$
$$^-5x = {}^-1000$$

$x = 200 \leftarrow$ the price of a cow was \$200.
Solving for y, $y = 150 \leftarrow$ the price of a sheep, \$150.

(This problem can also solved by substitution or determinants.)

Example: John has 9 coins, which are either dimes or nickels, that are worth \$.65. Determine how many of each coin he has.

Let $d =$ number of dimes.
Let $n =$ number of nickels.
The number of coins total 9.
The value of the coins equals 65.

Then: $n + d = 9$
$5n + 10d = 65$

Multiplying the first equation by $^-5$, it becomes:

$$^-5n - 5d = {}^-45$$
$$\underline{5n + 10d = 65}$$
$$5d = 20$$

$d = 4$ There are 4 dimes, so there are
$(9-4)$ or 5 nickels.

Example: Sharon's Bike Shoppe can assemble a 3 speed bike in 30 minutes or a 10 speed bike in 60 minutes. The profit on each bike sold is \$60 for a 3 speed or \$75 for a 10 speed bike. How many of each type of bike should they assemble during an 8 hour day (480 minutes) to make the maximum profit? Total daily profit must be at least \$300.

TEACHER CERTIFICATION EXAM

Let x = number of 3 speed bikes.
y = number of 10 speed bikes.

Since there are only 480 minutes to use each day,

$30x + 60y \leq 480$ is the first inequality.

Since the total daily profit must be at least \$300,

$60x + 75y \geq 300$ is the second inequality.

$32x + 65y \leq 480$ solves to $y \leq 8 - 1/2\,x$
$60x + 75y \geq 300$ solves to $y \geq 4 - 4/5\,x$

Graph these 2 inequalities:

$y \leq 8 - 1/2\,x$
$y \geq 4 - 4/5\,x$

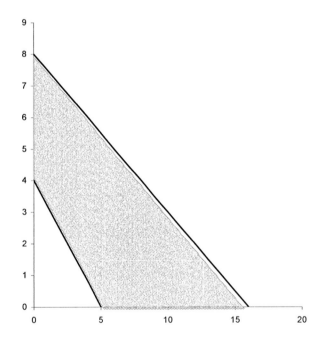

Realize that $x \geq 0$ and $y \geq 0$, since the number of bikes assembled can not be a negative number. Graph these as additional constraints on the problem. The number of bikes assembled must always be an integer value, so points within the shaded area of the graph must have integer values. The maximum profit will occur at or near a corner of the shaded portion of this graph. Those points occur at (0,4), (0,8), (16,0), or (5,0).
Since profits are \$60/3-speed or \$75/10-speed, the profit would be :

MATHEMATICS HIGH SCHOOL 108

(0,4) $60(0)+75(4)=300$
(0,8) $60(0)+75(8)=600$
(16,0) $60(16)+75(0)=960$ ← Maximum profit
(5,0) $60(5)+75(0)=300$

The maximum profit would occur if 16 3-speed bikes are made daily.

SKILL 16.7 Solve quadratic equations and inequalities.

To solve a quadratic equation(with x^2), rewrite the equation into the form:

$$ax^2 + bx + c = 0 \text{ or } y = ax^2 + bx + c$$

where a, b, and c are real numbers. Then substitute the values of a, b, and c into the quadratic formula:

$$x = \frac{-b \pm \sqrt{b^2 - 4ac}}{2a}$$

Simplify the result to find the answers. (Remember, there could be 2 real answers, one real answer, or 2 complex answers that include "i").

To solve a quadratic inequality (with x^2), solve for y. The axis of symmetry is located at $x = {}^-b/2a$. Find coordinates of points to each side of the axis of symmetry. Graph the parabola as a dotted line if the inequality sign is either < or >. Graph the parabola as a solid line if the inequality sign is either ≤ or ≥. Shade above the parabola if the sign is ≥ or >. Shade below the parabola if the sign is ≤ or <.

Example: Solve: $8x^2 - 10x - 3 = 0$

In this equation $a = 8$, $b = {}^-10$, and $c = {}^-3$.
Substituting these into the quadratic equation, it becomes:

$$x = \frac{-({}^-10) \pm \sqrt{({}^-10)^2 - 4(8)({}^-3)}}{2(8)} = \frac{10 \pm \sqrt{100 + 96}}{16}$$

$$x = \frac{10 \pm \sqrt{196}}{16} = \frac{10 \pm 14}{16} = 24/16 = 3/2 \text{ or } {}^-4/16 = {}^-1/4$$

Example: Solve and graph: $y > x^2 + 4x - 5$.

The axis of symmetry is located at $x = {}^-b/2a$. Substituting 4 for b, and 1 for a, this formula becomes:

$$x = {}^-(4)/2(1) = {}^-4/2 = {}^-2$$

Find coordinates of points to each side of $x = {}^-2$.

x	y
⁻5	0
⁻4	⁻5
⁻3	⁻8
⁻2	⁻9
⁻1	⁻8
0	⁻5
1	0

Graph these points to form a parabola. Draw it as a dotted line. Since a greater than sign is used, shade above and inside the parabola.

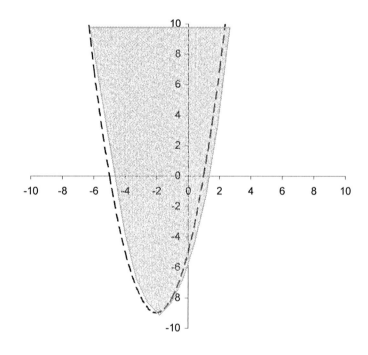

COMPETENCY 17.0   UNDERSTANDING OF POLYNOMIAL FUNCTIONS.

SKILL 17.1   Find the zeroes of a function and sketch the graph of a function using the factor theorem, the rational root theorem, the remainder theorem, synthetic division, or Descartes's rule of signs.

Synthetic division can be used to find the value of a function at any value of x. To do this, divide the value of x into the coefficients of the function. (Remember that coefficients of missing terms, like x^2 below, must be included.) The remainder of the synthetic division is the value of the function. If $f(x) = x^3 - 6x + 4$, then to find the value of the function at $x = 3$, use synthetic division:

Note the 0 for the missing x^2 term.

$$\begin{array}{r|rrrr} 3 & 1 & 0 & {}^-6 & 4 \\ & & 3 & 9 & 9 \\ \hline & 1 & 3 & 3 & 13 \end{array}$$ ← This is the value of the function.

Therefore, (3, 13) is a point on the graph.

Example: Find values of the function at $x = {}^-5$ if $f(x) = 2x^5 - 4x^3 + 3x^2 - 9x + 10$.

Note the 0 below for the missing x^4 term.

Synthetic division:

$$\begin{array}{r|rrrrrr} {}^-5 & 2 & 0 & {}^-4 & 3 & {}^-9 & 10 \\ & & {}^-10 & 50 & {}^-230 & 1135 & {}^-5630 \\ \hline & 2 & {}^-10 & 46 & {}^-227 & 1126 & {}^-5620 \end{array}$$

2 $^-$10 46 $^-$227 1126 $^-$5620 ← This is the value of the function if $x = {}^-5$.
Therefore, ($^-$5, $^-$5620) is a point on the graph.

Note that if $x = {}^-5$, the same value of the function can also be found by substituting $^-$5 in place of x in the function.

$$f({}^-5) = 2({}^-5)^5 - 4({}^-5)^3 + 3({}^-5)^2 - 9({}^-5) + 10$$
$$= 2({}^-3125) - 4({}^-125) + 3(25) - 9({}^-5) + 10$$
$$= {}^-6250 + 500 + 75 + 45 + 10 = {}^-5620$$

Therefore, ($^-$5, $^-$5620) is still a point on the graph.

To determine if $(x-a)$ or $(x+a)$ is a factor of a polynomial, do a synthetic division, dividing by the opposite of the number inside the parentheses. To see if $(x-5)$ is a factor of a polynomial, divide it by 5. If the remainder of the synthetic division is zero, then the binomial is a factor of the polynomial.

If $f(x) = x^3 - 6x + 4$, determine if $(x-1)$ is a factor of $f(x)$. Use synthetic division and divide by 1:

Note the 0 for the missing x^2 term.

$$
\begin{array}{r|rrrr}
1 & 1 & 0 & {}^-6 & 4 \\
  &   & 1 & 1 & {}^-5 \\
\hline
  & 1 & 1 & {}^-5 & {}^-1
\end{array}
$$
← This is the remainder of the function.

Therefore, $(x-1)$ is **not** a factor of $f(x)$.

If $f(x) = x^3 - 6x + 4$, determine if $(x-2)$ is a factor of $f(x)$. Use synthetic division and divide by 2:

$$
\begin{array}{r|rrrr}
2 & 1 & 0 & {}^-6 & 4 \\
  &   & 2 & 4 & {}^-4 \\
\hline
  & 1 & 2 & {}^-2 & 0
\end{array}
$$
← This is the remainder of the function.

Therefore, $(x-2)$ **is** a factor of $f(x)$.

The converse of this is also true. If you divide by k in any synthetic division and get a remainder of zero for the division, then $(x-k)$ is a factor of the polynomial. Similarly, if you divide by ^-k in any synthetic division and get a remainder of zero for the division, then $(x+k)$ is a factor of the polynomial.

Divide $2x^3 - 6x - 104$ by 4. What is your conclusion?

$$
\begin{array}{r|rrrr}
4 & 2 & 0 & {}^-6 & {}^-104 \\
  &   & 8 & 32 & 104 \\
\hline
  & 2 & 8 & 26 & 0
\end{array}
$$
← This is the remainder of the function.

Since the remainder is **0**, then **$(x-4)$** is a factor.

Given any polynomial, be sure that the exponents on the terms are in descending order. List out all of the factors of the first term's coefficient and of the constant in the last term. Make a list of fractions by putting each of the factors of the first term's coefficient over each of the factors of the last term. Reduce fractions when possible. Put a ± in front of each fraction. This list of fractions is a list of the only possible rational roots of a function. If the polynomial is of degree **n**, then at most n of these will actually be roots of the polynomial.

Example: List the possible rational roots for the function $f(x) = x^2 - 5x + 4$.

$$\pm \frac{\text{factors of 4}}{\text{factors of 1}} = \pm 1, 2, 4 \leftarrow 6 \text{ possible rational roots}$$

Example: List the possible rational roots for the function $f(x) = 6x^2 - 5x - 4$.

Make fractions of the form :

possible rational roots = $\pm \dfrac{\text{factors of 4}}{\text{factors of 6}} = \pm \dfrac{1,2,4}{1,2,3,6} =$
$\pm \dfrac{1}{2}, \dfrac{1}{3}, \dfrac{1}{6}, \dfrac{2}{3}, \dfrac{4}{3}, 1, 2, 4$ are the only 16 rational numbers that could be roots.

Since this equation is of degree 2, there are, at most, 2 rational roots. (They happen to be 4/3 and ⁻1/2.)

Descarte's Rule of signs can help to determine how many positive real roots or how many negative real roots a function would have. Given any polynomial, be sure that the exponents on the terms are in descending order. Count the number of successive terms of the polynomial where there is a sign change. The number of positive roots will be equal to the number of sign changes or will be less than the number of sign changes by a multiple of 2. For example,

$y = 2x^5 + 3x^4 - 6x^3 + 4x^2 + 8x - 9$ has 3 sign
$\quad\quad\quad\quad\quad\;\; \rightarrow 1 \quad\;\; \rightarrow 2 \quad\quad\;\;\; \rightarrow 3 \quad$ changes.

That means that this equation will have either 3 positive roots or 1 positive root.

For the equation:

$$y = 4x^6 - 5x^5 + 6x^4 - 3x^3 + 2x^2 + 8x - 10 \quad \text{has 5 sign}$$
$$\to 1 \to 2 \to 3 \to 4 \to 5 \quad \text{changes.}$$

This equation will have either 5 positive roots (equal to the number of sign changes) or 3 positive roots (2 less) or only 1 positive root(4 less).

For the equation:

$$y = x^8 - 3x^5 - 6x^4 - 3x^3 + 2x^2 - 8x + 10 \quad \text{has 4 sign}$$
$$\to 1 \to 2 \to 3 \to 4 \quad \text{changes.}$$

This equation will have either 4 positive roots (equal to the number of sign changes) or 2 positive roots (2 less) or no positive roots(4 less).

The second part of Descarte's Rule of signs also requires that terms are in descending order of exponents. Next look at the equation and change the signs of the terms that have an odd exponent. Then count the number of sign changes in successive terms of the new polynomial. The number of negative terms will be equal to the number of sign changes or will be less than the number of sign changes by a multiple of 2.

For example, given the equation:

$$y = 2x^5 + 3x^4 - 6x^3 + 4x^2 + 8x - 9$$

Change the signs of the terms with odd exponents.

$$y = {}^-2x^5 + 3x^4 + 6x^3 + 4x^2 - 8x - 9$$

Now count the number of sign changes in this equation.

$$y = {}^-2x^5 + 3x^4 + 6x^3 + 4x^2 - 8x - 9 \quad \text{has 2 sign}$$
$$\to 1 \to 2 \text{changes.}$$

This tells you that there are 2 negative roots or 0 negative roots (2 less).

Example: Determine the number of positive or negative real roots for the equation:

$$y = x^3 + 9x^2 + 23x + 15$$

This equation is of degree 3, so it has, at most, 3 roots. Look at the equation. There are 0 sign changes. This means there are 0 positive roots. To check for negative roots, change the signs of the terms with odd exponents. The equation becomes:

$$y = {}^-x^3 + 9x^2 - 23x + 15 \quad \text{Now count sign changes.}$$
$$\phantom{y = {}^-x^3 + 9x^2 - 23x + 15 \quad} \rightarrow 1 \quad \rightarrow 2 \quad \rightarrow 3 \quad \text{There are 3 sign changes.}$$

This means there are either 3 negative roots or only 1 negative root.

To find points on the graph of a polynomial, substitute desired values in place of x and solve for the corresponding y value of that point on the graph. A second way to do the same thing is to do a synthetic division, dividing by the x value of the desired point. The remainder at the end of the synthetic division is the y value of the point. Find a group of points to plot, graph and connect the points from left to right on the graph. The y intercept will always have a y value equal to the constant of the equation.

SKILL 17.2 Determine the inverse of a given function.

How to write the equation of the inverse of a function.

1. To find the inverse of an equation using x and y, replace each letter with the other letter. Then solve the new equation for y, when possible. Given an equation like $y = 3x - 4$, replace each letter with the other:

$x = 3y - 4$. Now solve this for y:
$x + 4 = 3y$
$1/3 x + 4/3 = y$ This is the inverse.

Sometimes the function is named by a letter:
$f(x) = 5x + 10$

Temporarily replace f(x) with y.
$y = 5x + 10$

Now replace each letter with the other: $x = 5y + 10$
Solve for the new y: $x - 10 = 5y$
$1/5 x - 2 = y$

The inverse of f(x) is denoted as $f^{-1}(x)$, so the answer is $f^{-1}(x) = 1/5 X - 2$.

COMPTETENCY 18.0 ABILITY TO SOLVE PROBLEMS INVOLVING RATIONAL EXPONENTS.

SKILL 18.1 Evaluate expressions with rational exponents.

When an expression has a rational exponent, it can be rewritten using a radical sign. The denominator of the rational exponent becomes the index number on the front of the radical sign. The base of the original expression goes inside the radical sign. The numerator of the rational exponent is an exponent which can be placed either inside the radical sign on the original base or outside the radical as an exponent on the radical expression. The radical can then be simplified as far as possible.

$$4^{3/2} = \sqrt[2]{4^3} \text{ or } \left(\sqrt{4}\right)^3 = \sqrt{64} = 8$$

$$16^{3/4} = \sqrt[4]{16^3} \text{ or } \left(\sqrt[4]{16}\right)^3 = 2^3 = 8$$

$$25^{-1/2} = \frac{1}{25^{1/2}} = \frac{1}{\sqrt{25}} = \frac{1}{5}$$

SKILL 18.2 Rewrite expressions involving radicals as expressions with rational number exponents.

An expression with a radical sign can be rewritten using a rational exponent. The radicand becomes the base which will have the rational exponent. The index number on the front of the radical sign becomes the denominator of the rational exponent. The numerator of the rational exponent is the exponent which was originally inside the radical sign on the original base. Note: If no index number appears on the front of the radical, then it is a 2. If no exponent appears inside the radical, then use a 1 as the numerator of the rational exponent.

$$\sqrt[5]{b^3} = b^{3/5}$$

$$\sqrt[4]{ab^3} = a^{1/4}b^{3/4}$$

SKILL 18.3 Rewrite expressions involving rational exponents as expressions with radicals.

When an expression has a rational exponent, it can be rewritten using a radical sign. The denominator of the rational exponent becomes the index number on the front of the radical sign. The base of the original expression goes inside the radical sign. The numerator of the rational exponent is an exponent which can be placed either inside the radical sign on the original base or outside the radical as an exponent on the radical expression.

$$a^{2/9}b^{4/9}c^{8/9} = \sqrt[9]{a^2 b^4 c^8}$$

$$3^{1/5} = \sqrt[5]{3}$$

If an expression contains rational expressions with different denominators, rewrite the exponents with a common denominator and then change the problem into a radical.

$$a^{2/3}b^{1/2}c^{3/5} = a^{20/30}b^{15/30}c^{18/30} = \sqrt[30]{a^{20}b^{15}c^{18}}$$

SKILL 18.4 Solve radical equations.

To solve a radical equation:

1. Get a term with a radical alone on one side of the equation.

2. Raise both sides of the equation to a power equal to the index on the radical (that is, square both sides of an equation containing a square root. Cube both sides of an equation containing a cube root etc.). DO NOT SQUARE (OR CUBE, etc.) each term separately. SQUARE (OR CUBE, etc.) the ENTIRE SIDE of the equation.

3. If there are any radicals remaining, repeat steps 1 and 2 until all radicals are gone.

4. Solve the remaining equation.

5. Check your answers in the original radical equation. Every answer may not check. If no answer checks in the original equation, then the answer to the equation is $\varnothing$, the empty set or null set.

TEACHER CERTIFICATION EXAM

NOTE: Since answers on this type of problem must be checked in the original equation anyhow, this means that possible answers on a multiple choice test could be immediately substituted into the problem to find the correct answer without having to solve the actual problem.

Solve and check:

$\sqrt{2x-8} - 7 = 9$ Get radical alone.

$\sqrt{2x-8} = 16$ Square both sides.

$\sqrt{2x-8}^2 = 16^2$

$2x - 8 = 256$ Solve for x.

$2x = 264$

$x = 132$

Check:

$\sqrt{2(132)-8} - 7 = 9$

$\sqrt{264-8} - 7 = 9$

$\sqrt{256} - 7 = 9$

$16 - 7 = 9$ This answer checks.

Solve and check:

$\sqrt{5x-1} - 1 = x$ Add 1.

$\left(\sqrt{5x-1}\right)^2 = (x+1)^2$ Square both sides.

$5x - 1 = x^2 + 2x + 1$ Solve this equation.

$0 = x^2 - 3x + 2$

$0 = (x-2)(x-1)$

$x = 2 \quad\quad x = 1$

Check both answers:

$\sqrt{5(2)-1} - 1 = 2$ $\sqrt{5(1)-1} - 1 = 1$

$\quad\quad 3 - 1 = 2$ $2 - 1 = 1$

When these are checked, both answers check.

MATHEMATICS HIGH SCHOOL

TEACHER CERTIFICATION EXAM

COMPETENCY 19.0   **PROFICIENCY IN PERFORMING OPERATIONS WITH COMPLEX NUMBERS.**

SKILL 19.1   Find the sum, difference, product, and quotient of two complex numbers.

Complex numbers are of the form $a + bi$, where a and b are real numbers and $i = \sqrt{-1}$. When i appears in an answer, it is acceptable unless it is in a denominator. When i^2 appears in a problem, it is always replaced by a $^-1$. Remember, $i^2 = {}^-1$.

To add or subtract complex numbers, add or subtract the real parts. Then add or subtract the imaginary parts and keep the i (just like combining like terms).

<u>Examples</u>:   Add $(2 + 3i) + ({}^-7 - 4i)$.

$2 + {}^-7 = {}^-5$ $3i + {}^-4i = {}^-i$ so,

$(2 + 3i) + ({}^-7 - 4i) = {}^-5 - i$

Subtract $(8 - 5i) - ({}^-3 + 7i)$
$8 - 5i + 3 - 7i = 11 - 12i$

To multiply 2 complex numbers, F.O.I.L. the 2 numbers together. Replace i^2 with a $^-1$ and finish combining like terms. Answers should have the form $a + bi$.

<u>Example</u>:   Multiply $(8 + 3i)(6 - 2i)$ F.O.I.L. this.
$48 - 16i + 18i - 6i^2$ Let $i^2 = {}^-1$.
$48 - 16i + 18i - 6({}^-1)$
$48 - 16i + 18i + 6$
$54 + 2i$ This is the answer.

<u>Example</u>:   Multiply $(5 + 8i)^2$ ← Write this out twice.
$(5 + 8i)(5 + 8i)$ F.O.I.L. this
$25 + 40i + 40i + 64i^2$ Let $i^2 = {}^-1$.
$25 + 40i + 40i + 64({}^-1)$
$25 + 40i + 40i - 64$
${}^-39 + 80i$ This is the answer.

When dividing 2 complex numbers, you must eliminate the complex number in the denominator. If the complex number in the denominator is of the form $b\,i$, multiply both the numerator and denominator by i. Remember to replace i^2 with a $^-1$ and then continue simplifying the fraction.

Example:

$$\frac{2+3i}{5i} \quad \text{Multiply this by } \frac{i}{i}$$

$$\frac{2+3i}{5i} \times \frac{i}{i} = \frac{(2+3i)\,i}{5i \cdot i} = \frac{2i+3i^2}{5i^2} = \frac{2i+3(^-1)}{^-5} = \frac{^-3+2i}{^-5} = \frac{3-2i}{5}$$

If the complex number in the denominator is of the form $a+b\,i$, multiply both the numerator and denominator by **the conjugate of the denominator**. **The conjugate of the denominator** is the same 2 terms with the opposite sign between the 2 terms (the real term does not change signs). The conjugate of $2-3i$ is $2+3i$. The conjugate of $^-6+11i$ is $^-6-11i$. Multiply together the factors on the top and bottom of the fraction. Remember to replace i^2 with a $^-1$, combine like terms, and then continue simplifying the fraction.

Example:

$$\frac{4+7i}{6-5i} \quad \text{Multiply by } \frac{6+5i}{6+5i}, \text{ the conjugate.}$$

$$\frac{(4+7i)}{(6-5i)} \times \frac{(6+5i)}{(6+5i)} = \frac{24+20i+42i+35i^2}{36+30i-30i+25i^2} = \frac{24+62i+35(^-1)}{36+25(^-1)} = \frac{^-11+62i}{11}$$

Example:

$$\frac{24}{^-3-5i} \quad \text{Multiply by } \frac{^-3+5i}{^-3+5i}, \text{ the conjugate.}$$

$$\frac{24}{^-3-5i} \times \frac{^-3+5i}{^-3+5i} = \frac{^-72+120i}{9-25i^2} = \frac{^-72+120i}{9+25} = \frac{^-72+120i}{34} = \frac{^-36+60i}{17}$$

Divided everything by 2.

COMPETENCY 20.0   ABILITY TO SOLVE PROBLEMS INVOLVING SEQUENCES AND SERIES.

SKILL 20.1   Find a specified term in an arithmetic sequence.

When given a set of numbers where the common difference between the terms is constant, use the following formula:

$$a_n = a_1 + (n-1)d$$ where a_1 = the first term
   n = the n th term (general term)
   d = the common difference

Sample problem:

1. Find the 8th term of the arithmetic sequence 5, 8, 11, 14, ...

   $a_n = a_1 + (n-1)d$
   $a_1 = 5$ Identify 1st term.
   $d = 3$ Find d.
   $a_8 = 5 + (8-1)3$ Substitute.
   $a_8 = 26$

2. Given two terms of an arithmetic sequence find a and d.

   $a_4 = 21$ $a_6 = 32$
   $a_n = a_1 + (n-1)d$
   $21 = a_1 + (4-1)d$
   $32 = a_1 + (6-1)d$

   $21 = a_1 + 3d$ Solve the system of equations.
   $32 = a_1 + 5d$

   $21 = a_1 + 3d$
   $-32 = {}^-a_1 - 5d$ Multiply by $^-1$ and add the equations.
   ${}^-11 = {}^-2d$
   $5.5 = d$

   $21 = a_1 + 3(5.5)$ Substitute $d = 5.5$ into one of the equations.
   $21 = a_1 + 16.5$
   $a_1 = 4.5$

MATHEMATICS HIGH SCHOOL

The sequence begins with 4.5 and has a common difference of 5.5 between numbers.

SKILL 20.2 Find a specified term in a geometric sequence.

When using geometric sequences consecutive numbers are compared to find the common ratio.

$$r = \frac{a_{n+1}}{a_n}$$

$r = $ the common ratio
$a_n = $ the nth term

The ratio is then used in the geometric sequence formula:

$$a_n = a_1 r^{n-1}$$

Sample problems:

1. Find the 8th term of the geometric sequence 2, 8, 32, 128 ...

$r = \dfrac{a_{n+1}}{a_n}$ Use the common ratio formula to find ratio.

$r = \dfrac{8}{2}$ Substitute $a_n = 2$ $a_{n+1} = 8$

$r = 4$

$a_n = a_1 \times r^{n-1}$ Use $r = 4$ to solve for the 8th term.
$a_8 = 2 \times 4^{8-1}$
$a_8 = 32768$

MATHEMATICS HIGH SCHOOL

TEACHER CERTIFICATION EXAM

SKILL 20.3 Determine the sum of terms in a progression.

The sums of terms in a progression is simply found by determining if it is an arithmetic or geometric sequence and then using the appropriate formula.

Sum of first n terms of an arithmetic sequence.

$$S_n = \frac{n}{2}(a_1 + a_n)$$

or

$$S_n = \frac{n}{2}\left[2a_1 + (n-1)d\right]$$

Sum of first n terms of a geometric sequence.

$$S_n = \frac{a_1(r^n - 1)}{r - 1}, r \neq 1$$

Sample Problems:

1. $\sum_{i=1}^{10}(2i + 2)$ This means find the sum of the terms beginning with the first term and ending with the 10th term of the sequence $a = 2i + 2$.

$a_1 = 2(1) + 2 = 4$
$a_{10} = 2(10) + 2 = 22$

$S_n = \frac{n}{2}(a_1 + a_n)$
$S_n = \frac{10}{2}(4 + 22)$
$S_n = 130$

2. Find the sum of the first 6 terms in an arithmetic sequence if the first term is 2 and the common difference is $^-3$.

$n = 6 \quad a_1 = 2 \quad d = {}^-3$

$S_n = \frac{n}{2}\left[2a_1 + (n-1)d\right]$

$S_6 = \frac{6}{2}\left[2 \times 2 + (6-1)^-3\right]$ Substitute known values.

$S_6 = 3\left[4 + (^-15)\right]$ Solve.

$S_6 = {}^-33$

MATHEMATICS HIGH SCHOOL

3. Find $\sum_{i=1}^{5} 4 \times 2^i$ This means the sum of the first 5 terms where $a_i = a \times b^i$ and $r = b$.

$a_1 = 4 \times 2^1 = 8$ Identify a_1, r, n
$r = 2 \quad n = 5$

$S_n = \dfrac{a_1(r^n - 1)}{r - 1}$ Substitute a, r, n

$S_5 = \dfrac{8(2^5 - 1)}{2 - 1}$ Solve.

$S_5 = \dfrac{8(31)}{1} = 248$

Practice problems:

1. Find the sum of the first five terms of the sequence if $a = 7$ and $d = 4$.

2. $\sum_{i=1}^{7} (2i - 4)$

2. Find the sum of the geometric sequence 2, 6, 18, 54, 162, 486, 1458…

4. $\sum_{i=1}^{6} -3\left(\dfrac{2}{5}\right)^i$

TEACHER CERTIFICATION EXAM

COMPETENCY 21.0   ABILITY TO SOLVE PROBLEMS INVOLVING PERMUTATIONS AND COMBINATIONS.

SKILL 21.1   Solve problems involving permutations and combinations.

The difference between permutations and combinations is that in permutations all possible ways of writing an arrangement of objects are given while in a combination a given arrangement of objects is listed only once.

Given the set {1, 2, 3, 4}, list the arrangements of two numbers that can be written as a combination and as a permutation.

Combination	Permutation
12, 13, 14, 23, 24, 34	12, 21, 13, 31, 14, 41,
	23, 32, 24, 42, 34, 43,
six ways	twelve ways

Using the formulas given below the same results can be found.

$$_nP_r = \frac{n!}{(n-r)!}$$
The notation $_nP_r$ is read "the number of permutations of n objects taken r at a time."

$$_4P_2 = \frac{4!}{(4-2)!}$$
Substitue known values.

$$_4P_2 = 12$$
Solve.

$$_nC_r = \frac{n!}{(n-r)!r!}$$
The number of combinations when r objects are selected from n objects.

$$_4C_2 = \frac{4!}{(4-2)!2!}$$
Substitute known values.

$$_4C_2 = 6$$
Solve.

TEACHER CERTIFICATION EXAM

COMPETENCY 22.0 KNOWLEDGE OF THE BINOMIAL THEOREM.

SKILL 22.1 Expand given binomials to a specified positive integral power.

The binomial expansion theorem is another method used to find the coefficients of $(x+y)$. Although Pascal's Triangle is easy to use for small values of n, it can become cumbersome to use with larger values of n.

Binomial Theorem:

For any positive value of n,

$$(x+y)^n = x^n + \frac{n!}{(n-1)!1!}x^{n-1}y + \frac{n!}{(n-2)!2!}x^{n-2}y^2 + \frac{n!}{(n-3)!3!}x^{n-3}y^3 + \frac{n!}{1!(n-1)!}xy^{n-1} + y^n$$

Sample Problem:

1. Expand $(3x+y)^5$

$$(3x)^5 + \frac{5!}{4!1!}(3x)^4 y^1 + \frac{5!}{3!2!}(3x)^3 y^2 + \frac{5!}{2!3!}(3x)^2 y^3 + \frac{5!}{1!4!}(3x)^1 y^4 + y^5 =$$

$$243x^5 + 405x^4 y + 270x^3 y^2 + 90x^2 y^3 + 15xy^4 + y^5$$

SKILL 22.2 Determine a specified term in the expansion of given binomials.

Any term of a binomial expansion can be written individually. For example, the y value of the seventh term of $(x+y)^n$, would be raised to the 6th power and since the sum of exponents on x and y must equal seven, then the x must be raised to the $n-6$ power.

The formula to find the rth term of a binomial expansion is:

$$\frac{n!}{[n-(r-1)]!(r-1)!}x^{n-(r-1)}y^{r-1}$$

where $r =$ the number of the desired term and $n =$ the power of the binomial.

MATHEMATICS HIGH SCHOOL

TEACHER CERTIFICATION EXAM

Sample Problem:

1. Find the third term of $(x+2y)^{11}$

$x^{n-(r-1)}$	y^{r-1}	Find x and y exponents.
$x^{11-(3-1)}$	y^{3-1}	
x^9	y^2	
$\dfrac{11!}{9!2!}(x^9)(2y)^2$		Substitute known values.
$110x^9y^2$		Solve.

Practice problems:

1. $(x+y)^7$; 5th term

2. $(3x-y)^9$; 3rd term

COMPETENCY 23.0 KNOWLEDGE OF THE FUNDAMENTAL GEOMETRIC CONCEPTS.

SKILL 23.1 Represent the relationships among points, lines, and planes, including the union and intersection of these sets.

In geometry the point, line and plane are key concepts and can be discussed in relation to each other.

collinear points   non-collinear points
are all on the same line are not on the same line

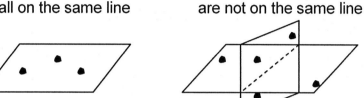

coplanar points   non-coplanar points
are on the same plane   are not on the same plane

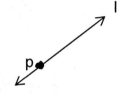

Point p is in line l
Point p is on line l
l contains P
l passes through P

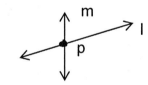

l and m intersect
at p
p is the intersection
of l and m

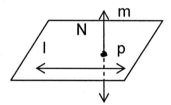

l and p are in plane N
N contains p and l
m intersects N at p
p is the intersection
of m and N

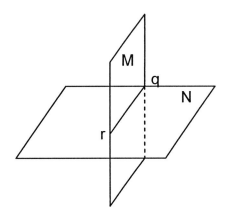

Planes M and N intersect at rq
rq is the intersection of M and N
rq is in M and N
M and N contain rq

SKILL 23.2 Label and classify angles according to their measures.

The classifying of angles refers to the angle measure. The naming of angles refers to the letters or numbers used to label the angle.

Sample Problem:

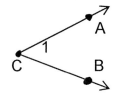

$\overrightarrow{CA}$ (read ray CA) and $\overrightarrow{CB}$ are the sides of the angle.
The angle can be called ∠ACB, ∠BCA, ∠C or ∠1.

Angles are classified according to their size as follows:

    acute: greater than 0 and less than 90 degrees.
    right: exactly 90 degrees.
    obtuse: greater than 90 and less than 180 degrees.
    straight: exactly 180 degrees

SKILL 23.3 Classify angle relationships (adjacent, complementary, supplementary, vertical, corresponding, alternate interior, or alternate exterior).

Angles can be classified in a number of ways. Some of those classifications are outlined here.

Adjacent angles have a common vertex and one common side but no interior points in common.

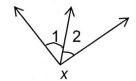

Complimentary angles add up to 90 degrees.

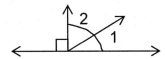

Supplementary angles add up to 180 degrees.

Vertical angels have sides that form two pairs of opposite rays.

Corresponding angles are in the same corresponding position on two parallel lines cut by a transversal.

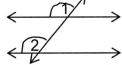

Alternate interior angles are diagonal angles on the inside of two parallel lines cut by a transversal.

Alternate exterior angles are diagonal on the outside of two parallel lines cut by a transversal.

SKILL 23.4 Classify lines and planes as perpendicular, intersecting parallel, or skew.

Parallel lines or planes do not intersect.

Perpendicular lines or planes form a 90 degree angle to each other.

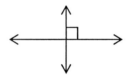

Intersecting lines share a common point and intersecting planes share a common set of points or line.

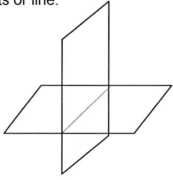

Skew lines do not intersect and do not lie on the same plane.

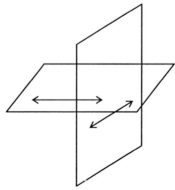

TEACHER CERTIFICATION EXAM

COMPETENCY 24.0   KNOWLEDGE OF TYPES OF POLYGONS AND THEIR ANGLES.

SKILL 24.1   Classify triangles by using the lengths of their sides and the measures of their angles.

A **triangle** is a polygon with three sides.

Triangles can be classified by the types of angles or the lengths of their sides.

Classifying by angles:

An **acute** triangle has exactly three *acute* angles.
A **right** triangle has one *right* angle.
An **obtuse** triangle has one *obtuse* angle.

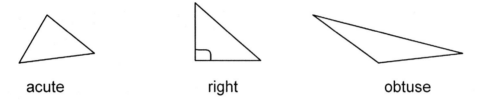

acute   right   obtuse

Classifying by sides:

All *three* sides of an **equilateral** triangle are the same length.
Two sides of an **isosceles** triangle are the same length.
None of the sides of a **scalene** triangle are the same length.

   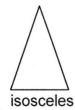   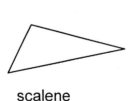

equilateral   isosceles   scalene

SKILL 24.2 Determine the measures of interior and exterior angles of a triangle, given appropriate information.

The sum of the measures of the angles of a triangle is 180°.

Example 1:
Can a triangle have two right angles?
No. A right angle measures 90°, therefore the sum of two right angles would be 180° and there could not be third angle.

Example 2:
Can a triangle have two obtuse angles?
No. Since an obtuse angle measures more than 90° the sum of two obtuse angles would be greater than 180°.

Example 3:
Can a right triangle be obtuse?
No. Once again, the sum of the angles would be more than 180°.

Example 4:
In a triangle, the measure of the second angle is three times the first. The third angle equals the sum of the measures of the first two angles. Find the number of degrees in
each angle.

Let x = the number of degrees in the first angle
$3x$ = the number of degrees in the second angle
$x + 3x$ = the measure of the third angle

Since the sum of the measures of all three angles is 180°.

$$x + 3x + (x + 3x) = 180$$
$$8x = 180$$
$$x = 22.5$$
$$3x = 67.5$$
$$x + 3x = 90$$

Thus the angles measure 22.5°, 67.5°, and 90°. Additionally, the triangle is a right triangle.

EXTERIOR ANGLES

Two adjacent angles form a linear pair when they have a common side and their remaining sides form a straight angle. Angles in a linear pair are supplementary. An exterior angle of a triangle forms a linear pair with an angle of the triangle.

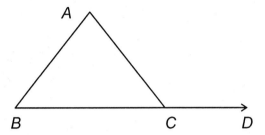

∠ACD is an exterior angle of triangle ABC, forming a linear pair with ∠ACB.

The measure of an exterior angle of a triangle is equal to the sum of the measures of the two non-adjacent interior angles.

Example:
In triangle ABC, the measure of ∠A is twice the measure of ∠B. ∠C is 30° more than their sum. Find the measure of the exterior angle formed at ∠C.

  Let x = the measure of ∠B
     $2x$ = the measure of ∠A
  $x + 2x + 30$ = the measure of ∠C
  $x + 2x + x + 2x + 30 = 180$
     $6x + 30 = 180$
       $6x = 150$
        $x = 25$
       $2x = 50$

It is not necessary to find the measure of the third angle, since the exterior angle equals the sum of the opposite interior angles. Thus the exterior angle at ∠C measures 75°.

SKILL 24.3 Determine the sum of the measures of the interior and exterior angles of convex polygons.

A **quadrilateral** is a polygon with four sides.
The sum of the measures of the angles of a quadrilateral is 360°.

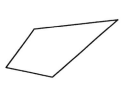 

A **trapezoid** is a quadrilateral with exactly <u>one</u> pair of parallel sides.

In an **isosceles trapezoid**, the non-parallel sides are congruent.

A **parallelogram** is a quadrilateral with <u>two</u> pairs of parallel sides.

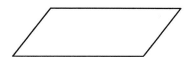

A **rectangle** is a parallelogram with a right angle.

A **rhombus** is a parallelogram with all sides equal length.

A **square** is a rectangle with all sides equal length.

SKILL 24.4 Determine the measures of the interior and exterior angles of a regular polygon, using the appropriate formula.

A **polygon** is a simple closed figure composed of line segments. In a **regular polygon** all sides are the same length and all angles are the same measure.

The sum of the measures of the **interior angles** of a polygon can be determined using the following formula, where n represents the number of angles in the polygon.

Sum of $\angle s = 180(n-2)$

The measure of each angle of a regular polygon can be found by dividing the sum of the measures by the number of angles.

Measure of $\angle = \dfrac{180(n-2)}{n}$

Example: Find the measure of each angle of a regular octagon.
Since an octagon has eight sides, each angle equals:

$$\dfrac{180(8-2)}{8} = \dfrac{180(6)}{8} = 135°$$

The sum of the measures of the **exterior angles** of a polygon, taken one angle at each vertex, equals 360°.

The measure of each exterior angle of a regular polygon can be determined using the following formula, where *n* represents the number of angles in the polygon.

Measure of exterior $\angle$ of regular polygon = $180 - \dfrac{180(n-2)}{n}$

or, more simply = $\dfrac{360}{n}$

Example: Find the measure of the interior and exterior angles of a regular pentagon.

Since a pentagon has five sides, each exterior angle measures:

$\dfrac{360}{5} = 72°$

Since each exterior angles is supplementary to its interior angle, the interior angle measures 180 - 72 or 108°.

SKILL 24.5 Define a parallelogram and apply related properties and theorems.

A **parallelogram** exhibits these properties.

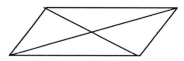

The diagonals bisect each other.
Each diagonal divides the parallelogram into two congruent triangles.
Both pairs of opposite sides are congruent.
Both pairs of opposite angles are congruent.
Two adjacent angles are supplementary.

Example 1:
Find the measures of the other three angles of a parallelogram if one angle measures 38°.

Since opposite angles are equal, there are two angles measuring 38°.
Since adjacent angles are supplementary, 180 - 38 = 142
so the other two angles measure 142° each.

```
      38
      38
     142
   + 142
     360
```

Example 2:
The measures of two adjacent angles of a parallelogram are $3x + 40$ and $x + 70$.

Find the measures of each angle.

$$2(3x + 40) + 2(x + 70) = 360$$
$$6x + 80 + 2x + 140 = 360$$
$$8x + 220 = 360$$
$$8x = 140$$
$$x = 17.5$$
$$3x + 40 = 92.5$$
$$x + 70 = 87.5$$

Thus the angles measure 92.5°, 92.5°, 87.5°, and 87.5°.

TEACHER CERTIFICATION EXAM

SKILL 24.6 Distinguish among special parallelograms.

Since a **rectangle** is a special type of parallelogram, it exhibits all the properties of a parallelogram. All the angles of a rectangle are right angles because of congruent opposite angles. Additionally, the diagonals of a rectangle are congruent.

A **rhombus** also has all the properties of a parallelogram. Additionally, its diagonals are perpendicular to each other and they bisect its angles.

A **square** has all the properties of a rectangle and a rhombus.

Example 1:
   True or false?

All squares are rhombuses.	True
All parallelograms are rectangles.	False - <u>some</u> parallelograms are rectangles
All rectangles are parallelograms.	True
Some rhombuses are squares.	True
Some rectangles are trapezoids.	False - only <u>one</u> pair of parallel sides
All quadrilaterals are parallelograms.	False - some quadrilaterals are parallelograms
Some squares are rectangles.	False - all squares are rectangles
Some parallelograms are rhombuses.	True

Example 2:
In rhombus $ABCD$ side $AB = 3x - 7$ and side $CD = x + 15$. Find the length of each side.
Since all the sides are the same length, $3x - 7 = x + 15$
$$2x = 22$$
$$x = 11$$
Since $3(11) - 7 = 25$ and $11 + 15 = 25$, each side measures 26 units.

MATHEMATICS HIGH SCHOOL

SKILL 24.7 Apply properties and theorems for trapezoids.

A **trapezoid** is a quadrilateral with exactly <u>one</u> pair of parallel sides.

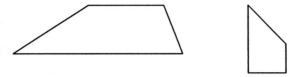

In an **isosceles trapezoid**, the non-parallel sides are congruent.

An isosceles trapezoid has the following properties:

The diagonals of an isosceles trapezoid are congruent.
The base angles of an isosceles trapezoid are congruent.

Example:

An isosceles trapezoid has a diagonal of 10 and a base angle measure of $30°$. Find the measure of the other 3 angles.

Based on the properties of trapezoids, the measure of the other base angle is $30°$ and the measure of the other diagonal is 10.

The other two angles have a measure of:

$$360 = 30(2) + 2x$$
$$x = 150°$$

The other two angles measure $150°$ each.

TEACHER CERTIFICATION EXAM

COMPETENCY 25.0   ABILITY TO INVESTIGATE AND APPLY THE CONCEPTS OF CONGRUENCE.

SKILL 25.1   Identify congruent polygons and their corresponding parts.

Congruent figures have the same size and shape. If one is placed above the other, it will fit exactly. Congruent lines have the same length. Congruent angles have equal measures.
The symbol for congruent is $\cong$.

Polygons (pentagons) *ABCDE* and *VWXYZ* are congruent. They are exactly the same size and shape.

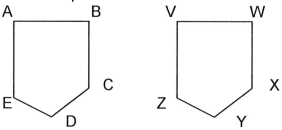

$ABCDE \cong VWXYZ$

Corresponding parts are those congruent angles and congruent sides, that is:

corresponding angles	corresponding sides
$\angle A \leftrightarrow \angle V$	$AB \leftrightarrow VW$
$\angle B \leftrightarrow \angle W$	$BC \leftrightarrow WX$
$\angle C \leftrightarrow \angle X$	$CD \leftrightarrow XY$
$\angle D \leftrightarrow \angle Y$	$DE \leftrightarrow YZ$
$\angle E \leftrightarrow \angle Z$	$AE \leftrightarrow VZ$

SKILL 25.2 Use the SAS, ASA, and SSS postulates to show pairs of triangles conguent.

Two triangles can be proven congruent by comparing pairs of appropriate congruent corresponding parts.

SSS POSTULATE

If three sides of one triangle are congruent to three sides of another triangle, then the two triangles are congruent.

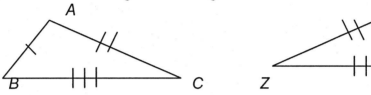

Since $AB \cong XY$, $BC \cong YZ$ and $AC \cong XZ$, then $\triangle ABC \cong \triangle XYZ$.

Example: Given isosceles triangle ABC with D the midpoint of base AC, prove the two triangles formed by AD are congruent.

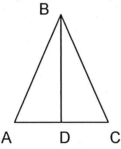

Proof:
1. Isosceles triangle ABC,
   D midpoint of base AC Given
2. $AB \cong BC$ An isosceles $\triangle$ has two congruent sides
3. $AD \cong DC$ Midpoint divides a line into two equal parts
4. $BD \cong BD$ Reflexive
5. $\triangle ABD \cong \triangle BCD$ SSS

SAS POSTULATE

If two sides and the included angle of one triangle are congruent to two sides and the included angle of another triangle, then the two triangles are congruent.

Example:

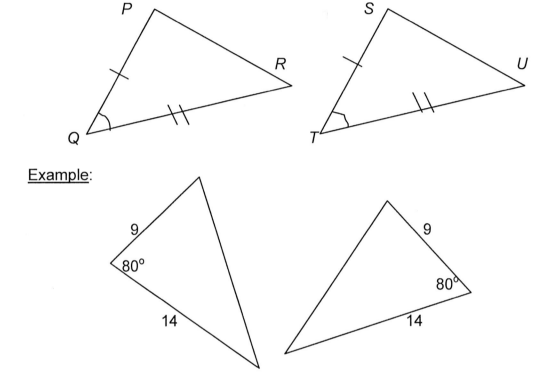

The two triangles are congruent by SAS.

ASA POSTULATE

If two angles and the included side of one triangle are congruent to two angles and the included side of another triangle, the triangles are congruent.

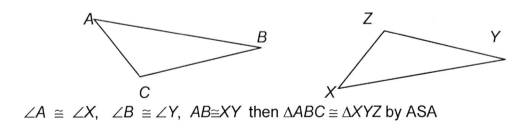

$\angle A \cong \angle X$, $\angle B \cong \angle Y$, $AB \cong XY$ then $\triangle ABC \cong \triangle XYZ$ by ASA

Example 1: Given two right triangles with one leg of each measuring 6 cm and the adjacent angle 37°, prove the triangles are congruent.

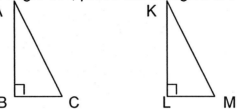

1. Right triangles ABC and KLM   Given
   AB = KL = 6 cm
   ∠A = ∠K = 37°
2. AB ≅ KL   Figures with the same
   ∠A ≅ ∠K   measure are congruent
3. ∠B ≅ ∠L   All right angles are
              congruent.
4. △ABC ≅ △KLM   ASA

Example 2:
What method would you use to prove the triangles congruent?

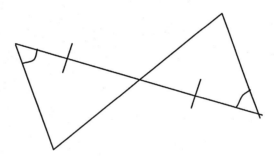

ASA because vertical angles are congruent.

MATHEMATICS HIGH SCHOOL

SKILL 25.3 Use the AAS and HL theorums to show pairs of triangles congruent.

AAS THEOREM

If two angles and a non-included side of one triangle are congruent to the corresponding parts of another triangle, then the triangles are congruent.

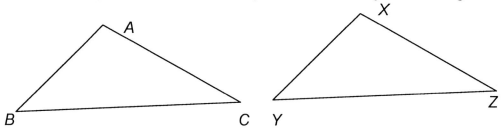

∠B ≅ ∠Y, ∠C ≅ ∠Z, AC ≅ XZ, then △ABC ≅ △XYZ by AAS.

We can derive this theorem because if two angles of the triangles are congruent, then the third angle must also be congruent. Therefore, we can uses the ASA postulate.

HL THEOREM

If the hypotenuse and a leg of one right triangle are congruent to the corresponding parts of another right triangle, the triangles are congruent.

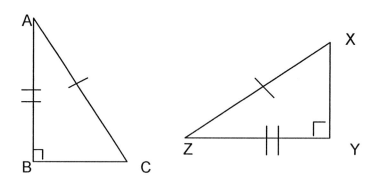

Since ∠B and ∠Y are right angles and AC ≅ XZ (hypotenuse of each triangle), AB ≅ YZ (corresponding leg of each triangle), then △ABC ≅△XYZ by HL.

TEACHER CERTIFICATION EXAM

<u>Example</u>: What method would you use to prove the triangles congruent?

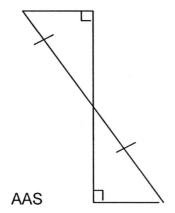

AAS

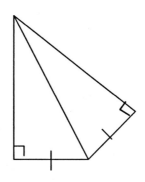

HL

COMPETENCY 26.0   ABILITY TO APPLY THE PROPERTIES THAT RELATE TO SIMILAR POLYGONS.

SKILL 26.1   Define similar polygons and apply related properties and theorems.

Two figures that have the **same shape** are **similar**. Two polygons are similar if corresponding angles are congruent and corresponding sides are in proportion. Corresponding parts of similar polygons are proportional.

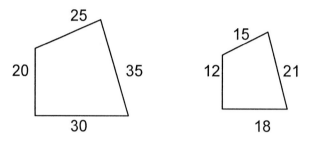

SIMILAR TRIANGLES

AA Similarity Postulate
If two angles of one triangle are congruent to two angles of another triangle, then the triangles are similar.

SAS Similarity Theorem
If an angle of one triangle is congruent to an angle of another triangle and the sides adjacent to those angles are in proportion, then the triangles are similar.

SSS Similarity Theorem
If the sides of two triangles are in proportion, then the triangles are similar.

SKILL 26.2 Use ratios and proportions to solve problems.

Explanation can be found in Skill 26.1.

Example:

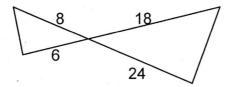

The two triangles are similar since the sides are proportional and vertical angles are congruent.

Example: Given two similar quadrilaterals. Find the lengths of sides *x*, *y*, and *z*.

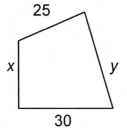 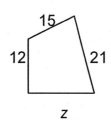

Since corresponding sides are proportional:

$$\frac{15}{25} = \frac{3}{5} \text{ so the scale is } \frac{3}{5}$$

$\dfrac{12}{x} = \dfrac{3}{5}$ $\dfrac{21}{y} = \dfrac{3}{5}$ $\dfrac{z}{30} = \dfrac{3}{5}$

$3x = 60$ $3y = 105$ $5z = 90$
$x = 20$ $y = 35$ $z = 18$

TEACHER CERTIFICATION EXAM

COMPETENCY 27.0   **ABILITY TO APPLY RELATIONSHIPS THAT EXIST IN RIGHT TRIANGLES.**

SKILL 27.1   **State and apply the relationships that exist when the altitude is drawn to the hypotenuse of a right triangle.**

A **right triangle** is a triangle with one right angle. The side opposite the right angle is called the **hypotenuse**. The other two sides are the **legs**. An **altitude** is a line drawn from one vertex, perpendicular to the opposite side.

When an altitude is drawn to the hypotenuse of a right triangle, then the two triangles formed are similar to the original triangle and to each other.

Example:

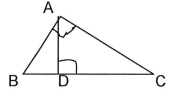

Given right triangle ABC with right angle at A,
altitude AD drawn to hypotenuse BC at D.

△ABC ~ △ABD ~ △ACD The triangles formed are similar to each other and to the original right triangle.

SKILL 27.2   **Find the geometric mean between two numbers.**

If a, b and c are positive numbers such that $\dfrac{a}{b} = \dfrac{b}{c}$
then b is called the **geometric mean** between a and c.

Example:

Find the geometric mean between 6 and 20.

$$\frac{6}{x} = \frac{x}{30}$$
$$x^2 = 180$$
$$x = \sqrt{180} = \sqrt{36 \cdot 5} = 6\sqrt{5}$$

The geometric mean is significant when the altitude is drawn to the hypotenuse of a right triangle.
The length of the altitude is the geometric mean between each segment of the hypotenuse,
   and
Each leg is the geometric mean between the hypotenuse and the segment of the hypotenuse that is adjacent to the leg.

Example:

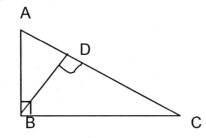

△ABC is a right △
BD is the altitude of △ABC
AB = 6
AC = 12
Find AD, CD, BD, and BC

$\dfrac{12}{6} = \dfrac{6}{AD}$ $\dfrac{3}{BD} = \dfrac{BD}{9}$ $\dfrac{12}{BC} = \dfrac{BC}{9}$

$12(AD) = 36$ $(BD)^2 = 27$ $(BC)^2 = 108$

$AD = 3$ $BD = \sqrt{27} = \sqrt{9 \cdot 3} = 3\sqrt{3}$ $BC = \sqrt{108} = \sqrt{36 \cdot 3} = 6\sqrt{3}$

$CD = 12 - 3 = 9$

TEACHER CERTIFICATION EXAM

SKILL 27.3 **Apply the Pythagorean theorem to obtain the lengths of the sides of right triangles.**

Pythagorean theorem states that the square of the length of the hypotenuse is equal to the sum of the squares of the lengths of the legs. Symbolically, this is stated as:

$$c^2 = a^2 + b^2$$

Given the right triangle below, find the missing side.

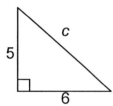

$c^2 = a^2 + b^2$ 1. write formula
$c^2 = 5^2 + 6^2$ 2. substitute known values
$c^2 = 61$ 3. take square root
$c = \sqrt{61}$ or 7.81 4. solve

SKILL 27.4 **Apply converse of the Pythagorean theorem and related theorems.**

The Converse of the Pythagorean Theorem states that if the square of one side of a triangle is equal to the sum of the squares of the other two sides, then the triangle is a right triangle.

Example:
Given $\triangle XYZ$, with sides measuring 12, 16 and 20 cm. Is this a right triangle?

$c^2 = a^2 + b^2$
20^2 ? $12^2 + 16^2$
400 ? $144 + 256$
$400 = 400$

Yes, the triangle is a right triangle.

This theorem can be expanded to determine if triangles are obtuse or acute.

MATHEMATICS HIGH SCHOOL

If the square of the longest side of a triangle is greater than the sum of the squares of the other two sides, then the triangle is an obtuse triangle.
and
If the square of the longest side of a triangle is less than the sum of the squares of the other two sides, then the triangle is an acute triangle.

Example:
Given $\triangle LMN$ with sides measuring 7, 12, and 14 inches. Is the triangle right, acute, or obtuse?

$14^2 \; ? \; 7^2 + 12^2$
$196 \; ? \; 49 + 144$
$196 > 193$

Therefore, the triangle is obtuse.

SKILL 27.5 Use the 30-60-90 and 45-45-90 triangle relationships to determine the lengths of the sides of appropriate triangles.

Given the special right triangles below, we can find the lengths of other special right triangles.

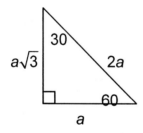 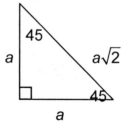

Sample problems:

1. 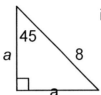 if $8 = a\sqrt{2}$ then $a = 8/\sqrt{2}$ or 5.657

2.   if $7 = a$ then $c = a\sqrt{2} = 7\sqrt{2}$ or 9.899

3. 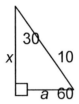 if $2a = 10$ then $a = 5$ and $x = a\sqrt{3} = 5\sqrt{3}$ or 8.66

SKILL 27.6 Define tangent, sine, and cosine ratios for acute angles.

Given triangle right ABC, the adjacent side and opposite side can be identified for each angle A and B.

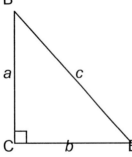

Looking at angle A, it can be determined that side *b* is adjacent to angle A and side *a* is opposite angle A.

If we now look at angle B, we see that side *a* is adjacent to angle *b* and side *b* is opposite angle B.

The longest side (opposite the 90 degree angle) is always called the hypotenuse.

The basic trigonometric ratios are listed below:

Sine = opposite/hypotenuse   Cosine = adjacent/hypotenuse   Tangent = opposite/adjacent

Sample problem:

1. Use triangle ABC to find the sin, cos and tan for angle A.

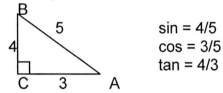

sin = 4/5
cos = 3/5
tan = 4/3

SKILL 27.7 Solve right triangle problems by correct selection and use of tangent, sine, and cosine ratios.

Use the basic trigonometric ratios of sine, cosine and tangent to solve for the missing sides of right triangles when given at least one of the acute angles.

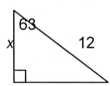

In the triangle ABC, an acute angle of 63 degrees and the length of the hypotenuse (12). The missing side is the one adjacent to the given angle.

The appropriate trigonometric ratio to use would be cosine since we are looking for the adjacent side and we have the length of the hypotenuse.

$\cos x$ = adjacent/hypotenuse   1. Write formula.

$\cos 63 = \dfrac{x}{12}$ 2. Substitute known values.

$0.454 = \dfrac{x}{12}$

$x = 5.448$ 3. Solve.

MATHEMATICS HIGH SCHOOL

TEACHER CERTIFICATION EXAM

COMPETENCY 28.0   **ABILITY TO APPLY PROPERTIES OF ANGLES AND LINES THAT ARE APPROPRIATE TO CIRCLES.**

SKILL 28.1   Calculate the degree measure of major and minor arcs, semicircles, and central angles.

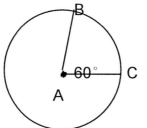

Central angle BAC = 60°
Minor arc BC = 60°
Major arc BC = 360 − 60 = 300°

If you draw two radii in a circle, the angle they form with the center as the vertex is a central angle. The piece of the circle "inside" the angle is an arc. Just like a central angle, an arc can have any degree measure from 0 to 360. The measure of an arc is equal to the measure of the central angle which forms the arc. Since a diameter forms a semicircle and the measure of a straight angle like a diameter is 180°, the measure of a semicircle is also 180°.

Given two points on a circle, there are two different arcs which the two points form. Except in the case of semicircles, one of the two arcs will always be greater than 180° and the other will be less than 180°. The arc less than 180° is a minor arc and the arc greater than 180° is a major arc.

Examples:

1.

    $m\angle BAD = 45°$
    What is the measure of the major arc BD?

∡BAD = minor arc BD   The measure of the central angle
45° = minor arc BD   is the same as the measure of
                                       the arc it forms.
360 − 45 = major arc BD   A major and minor arc always
315° = major arc BD   add to 360°.

2.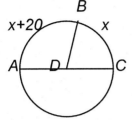

$\overline{AC}$ is a diameter of circle D. What is the measure of $\angle BDC$?

$m\angle ADB + m\angle BDC = 180°$ A diameter forms a semicircle
$x + 20 + x = 180$ which has a measure of $180°$.
$2x + 20 = 180$
$2x = 160$
$x = 80$
minor arc $BC = 80°$ A central angle has the same
$m\angle BDC = 80°$ measure as the arc it forms.

SKILL 28.2 Calculate the length of an arc and the area of sectors of a circle.

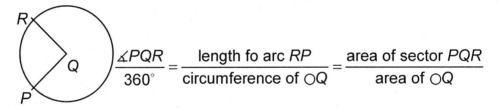

While an arc has a measure associated to the degree measure of a central angle, it also has a length which is a fraction of the circumference of the circle.

For each central angle and its associated arc, there is a sector of the circle which resembles a pie piece. The area of such a sector is a fraction of the area of the circle.

The fractions used for the area of a sector and length of its associated arc are both equal to the ratio of the central angle to $360°$.

Examples:

1.

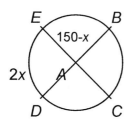

⊙A has a radius of 4 cm. What is the length of arc ED?

$2x + 150 - x = 180$
$x + 150 = 180$
$x = 30$
Arc $ED = 2(30) = 60°$
$\dfrac{60}{360} = \dfrac{\text{arc length } ED}{2\pi 4}$

$\dfrac{1}{6} = \dfrac{\text{arc length}}{8\pi}$

$\dfrac{8\pi}{6} = \text{arc length}$

arc length $ED = \dfrac{4\pi}{3}$ cm.

Arc BE and arc DE make a semicircle.

The ratio 60° to 360° is equal to the ratio of arch length ED to the circumference of ⊙A.

Cross multiply and solve for the arc length.

2.

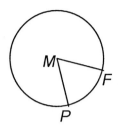

The radius of ⊙M is 3 cm. The length of arc PF is 2π cm. What is the area of sector PMF?

Circumference of ⊙$M = 2\pi(3) = 6\pi$ Find the circumference
Area of ⊙$M = \pi(3)^2 = 9\pi$ and area of the circle.

$\dfrac{\text{area of } PMF}{9\pi} = \dfrac{2\pi}{6\pi}$ The ratio of the sector area to the circle

$\dfrac{\text{area of } PMF}{9\pi} = \dfrac{1}{3}$ area is the same as the arc length to the

area of $PMF = \dfrac{9\pi}{3}$ circumference.

area of $PMF = 3\pi$ Solve for the area of the sector.

SKILL 28.3 Apply the theorems pertaining to the relationships of chords, diameters, radii, and tangents with respect to circles and to each other.

A tangent line intersects a circle in exactly one point. If a radius is drawn to that point, the radius will be perpendicular to the tangent.

A chord is a segment with endpoints on the circle. If a radius or diameter is perpendicular to a chord, the radius will cut the chord into two equal parts.

If two chords in the same circle have the same length, the two chords will have arcs that are the same length, and the two chords will be equidistant from the center of the circle. Distance from the center to a chord is measured by finding the length of a segment from the center perpendicular to the chord.

Examples:

1.

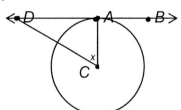

$\overrightarrow{DB}$ is tangent to $\odot C$ at A.
$m\angle ADC = 40°$. Find x.

$\overline{AC} \perp \overrightarrow{DB}$ — A radius is $\perp$ to a tangent at the point of tangency.

$m\angle DAC = 90°$ — Two segments that are $\perp$ form a $90°$ angle.

$40 + 90 + x = 180$ — The sum of the angles of a triangle is $180°$.

$x = 50°$ — Solve for x.

2.

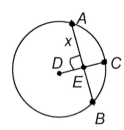

$\overline{CD}$ is a radius and $\overline{CD} \perp$ chord $\overline{AB}$.
$\overline{AB} = 10$. Find x.

$x = \dfrac{1}{2}(10)$

$x = 5$

If a radius is $\perp$ to a chord, the radius bisects the chord.

SKILL 28.4 Angles related to circles.

Angles with their vertices on the circle:

An inscribed angle is an angle whose vertex is on the circle. Such an angle could be formed by two chords, two diameters, two secants, or a secant and a tangent. An inscribed angle has one arc of the circle in its interior. The measure of the inscribed angle is one-half the measure of this intercepted arc. If two inscribed angles intercept the same arc, the two angles are congruent (i.e. their measures are equal). If an inscribed angle intercepts an entire semicircle, the angle is a right angle.

Angles with their vertices in a circle's interior:

When two chords intersect inside a circle, two sets of vertical angles are formed. Each set of vertical angles intercepts two arcs which are across from each other. The measure of an angle formed by two chords in a circle is equal to one-half the sum of the the angle intercepted by the angle and the arc intercepted by its vertical angle.

Angles with their veritices in a circle's exterior:

If an angle has its vertex outside of the circle and each side of the circle intersects the circle, then the angle contains two different arcs. The measure of the angle is equal to one-half the difference of the two arcs.

Examples:

1.

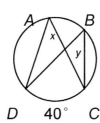

Find x and y.
arc $DC = 40°$

$m\angle DAC = \dfrac{1}{2}(40) = 20°$ $\angle DAC$ and $\angle DBC$ are both

$m\angle DBC = \dfrac{1}{2}(40) = 20°$ inscribed angles, so each one

$x = 20°$ and $y = 20°$ has a measure equal to one-half the measure of arc DC.

SKILL 28.5 Lengths of chords, secants, and tangents

Intersecting chords:

If two chords intersect inside a circle, each chord is divided into two smaller segments. The product of the lengths of the two segments formed from one chord equals the product of the lengths of the two segments formed from the other chord.

Intersecting tangent segments:

If two tangent segments intersect outside of a circle, the two segments have the same length.

Intersecting secant segments:

If two secant segments intersect outside a circle, a portion of each segment will lie inside the circle and a portion (called the exterior segment) will lie outside the circle. The product of the length of one secant segment and the length of its exterior segment equals the product of the length of the other secant segment and the length of its exterior segment.

Tangent segments intersecting secant segments:

If a tangent segment and a secant segment intersect outside a circle, the square of the length of the tangent segment equals the product of the length of the secant segment and its exterior segment.

Examples:

1.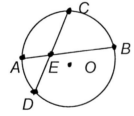

$\overline{AB}$ and $\overline{CD}$ are chords.
CE=10, ED=x, AE=5, EB=4

$(AE)(EB) = (CE)(ED)$
$5(4) = 10x$
$20 = 10x$
$x = 2$

Since the chords intersect in the circle, the products of the segment pieces are equal.
Solve for x.

2.

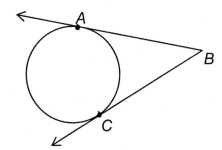

$\overline{AB}$ and $\overline{CD}$ are chords.
$\overline{AB} = x^2 + x - 2$
$\overline{BC} = x^2 - 3x + 5$
Find the length of $\overline{AB}$ and $\overline{BC}$.

$\overline{AB} = x^2 + x - 2$ $\overline{BC} = x^2 - 3x + 5$	Given
$\overline{AB} = \overline{BC}$	Intersecting tangents are equal.
$x^2 + x - 2 = x^2 - 3x + 5$	Set the expression equal and solve.
$4x = 7$ $x = 1.75$	Substitute and solve.
$(1.75)^2 + 1.75 - 2 = \overline{AB}$ $\overline{AB} = \overline{BC} = 2.81$	

MATHEMATICS HIGH SCHOOL

TEACHER CERTIFICATION EXAM

COMPETENCY 29.0   ABILITY TO PERFORM BASIC CONSTRUCTIONS WITH A COMPASS AND STRAIGHT EDGE.

SKILL 29.1   Construct a line segment congruent to a given line segment.

A geometric construction is a drawing made using only a compass and straightedge. A construction consists of only segments, arcs, and points. The easiest construction to make is to duplicate a given line segment. Given segment *AB*, construct a segment equal in length to segment AB by following these steps.

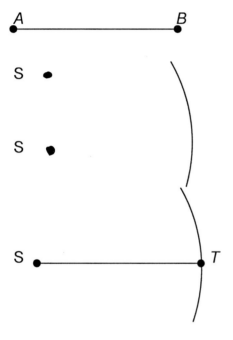

1. Place a point anywhere in the plane to Anchor the duplicate segment. Call this point S.

2. Open the compass to match the length of segment *AB*. Keeping the compass rigid, swing an arc from S.

3. Draw a segment from S to any point on the arc. This segment will be the same length as *AB*.

Samples:

Construct segments congruent to the given segments.

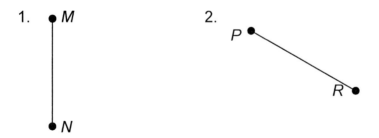

MATHEMATICS HIGH SCHOOL   164

TEACHER CERTIFICATION EXAM

SKILL 29.2 Construct an angle congruent to a given angle.

To construct an angle congruent to a given angle such as angle TAP follow these steps.

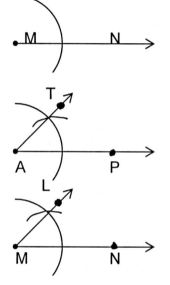

1. Draw ray *MN* using a straightedge. This ray will be one side of the duplicate angle.

2. Using the compass, draw an arc of any radius with its central at the vertex A. Draw an arc of the same radius with center M.

3. Use the point where the arc intercepts ray *AP* to draw another arc that intercepts the intersection of the arc and ray *AT*. Swing an arc of the same radius from the intersection point on ray *MN*.

4. Connect M and the point of intersection of the two arcs to form angle *LMN* which will be congruent to angle *TAP*.

MATHEMATICS HIGH SCHOOL

SKILL 29.3 Bisect an angle.

To bisect a given angle such as angle *FUZ*, follow these steps.

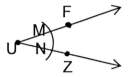

1. Swing an arc of any length with its center at point U. This arc will intersect rays *UF* and *UZ* at M and N.

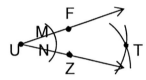

2. Open the compass to any length and swing one arc from point M and another arc of the same radius from point N. These arcs will intersect in the interior of angle *FUZ* at point T.

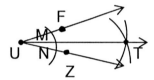

3. Connect U and T for the ray which bisects angle *FUZ*. Ray *UT* is the angle bisector of angle *FUZ*

SKILL 29.4 Given a point on a line, construct a perpendicular to the line through the point.

Given a line such as line $\overline{AB}$ and a point K on the line, follow these steps to construct a perpendicular line to line l through K.

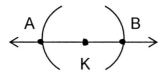

1. Swing an arc of any radius from point K so that it intersects line $\overline{AB}$ in two points, A and B.

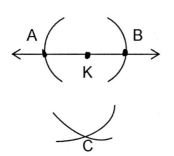

2. Open the compass to any length and swing one arc from B and another from A so that the two arcs intersect at point C.

3. Connect K and C to form line KC which is is perpendicular to line $\overline{AB}$.

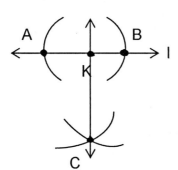

SKILL 29.5 Construct the perpendicular to a given line through a point from a given point not on the line.

Given a line such as line l and a point P not on l, follow these steps to construct a perpendicular line to l that passes through P.

1. Swing an arc of any radius from P so that the arc intersects line l in two points A and B.

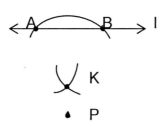

2. Open the compass to any length and swing two arcs of the same radius, one from A and the other from B. These two arcs will intersect at a new point K.

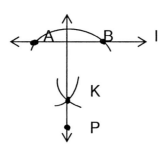

3. Connect K and P to form a line perpendicular to line l which passes through P.

TEACHER CERTIFICATION EXAM

SKILL 29.6 Construct the perpendicular bisector of a line segment of a given line segment.

Given a line segment with two endpoints such as A and B, follow these steps to construct the line which both bisects and is perpendicular to the line given segment.

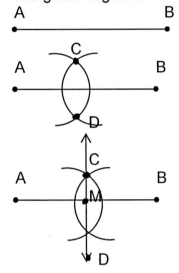

1. Swing an arc of any radius from point A. Swing another arc of the same radius from B. The arcs will intersect at two points. Label these points C and D.

2. Connect C and D to form the perpendicular bisector of segment *AB*.

3. The point M where line $\overline{CD}$ and segment $\overline{AB}$ intersect is the midpoint of segment $\overline{AB}$.

MATHEMATICS HIGH SCHOOL

SKILL 29.7 Construct a parallel to a given line through a given point not on the line.

Given a point such as P and a line such as line m, follow these steps to construct the single line which passes through P and is also parallel to line m.

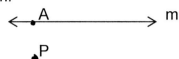

1. Place a point A anywhere on line m.

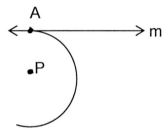

2. Open the compass to the distance between A and P. Using P as the center, swing a long Arc that passes through A.

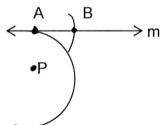

3. With the compass still open to the same length, swing an arc from A that intersects line m at a point labeled B.

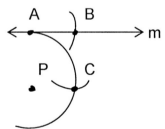

4. Using point B as the center, swing a new arc with the same radius as the other two arcs so that is intersects the arc from P at a new point C.

5. Connect P and C to obtain a line parallel to obtain a line parallel to line m

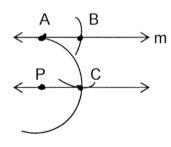

MATHEMATICS HIGH SCHOOL

TEACHER CERTIFICATION EXAM

SKILL 29.8 Construct the tangent to a circle at a given point.

Given a circle with center O and a point on the circle such as P, construct the line tangent to the circle at P by constructing the line perpendicular to the radius drawn to P. If a line is perpendicular to a radius, the line will be tangent to the circle. For constructing a tangent to a circle, follow these steps.

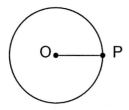

1. Draw the radius from O to P.

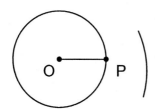

2. Open the compass from point P to point O. Use this radius to swing an arc from P to the exterior of the circle.

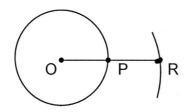

3. Put the straightedge on the radius and extend the segment to the arc forming segment OR. Note that P is the midpoint of OR.

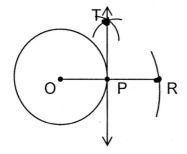

4. Open the compass to any length longer than the radius of this circle. Swing an arc of this radius from each endpoint of segment OR. Label the point where these two arcs intersect T.

5. Connect the point P and T to form the tangent line to circle O at point P.

SKILL 29.9 Construct an inscribed circle inside a given triangle.

For every triangle there is one circle which is inscribed inside the triangle, i.e. the three sides of the triangle are tangent to the circle. The center of the inscribed circle is the point where the angle bisectors of the triangle meet. The three angle bisectors meet in a single point of concurrency called the incenter. Since the three angle bisectors meet in a single point, you only need construct two of the angle bisectors to find the incenter. Follow these steps to find the inscribed circle for triangle ABC.

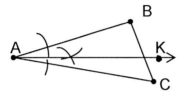

1. Construct the angle bisector of any angle. For constructing an angle bisector, see section 29.3. Label a point on the angle bisector point K, forming ray *AK*.

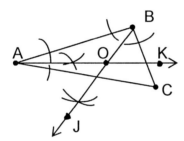

2. Construct the angle bisector of another angle, for example, angle B. Label a point on this bisector point J, forming ray *BJ*.

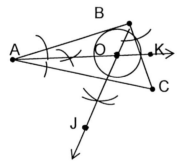

3. Label the point where the two angle bisectors meet point O. O is in the center of triangle ABC.

4. Place one end of the compass at point O, and open the compass so that it stretches to one of the triangle sides. Verify that if the compass is held rigid it will just reach to the other two sides. If it will not reach to the other sides, readjust the compass until you find the one length that will reach all three sides. With this radius, swing a circle around point O.

TEACHER CERTIFICATION EXAM

COMPETENCY 30.0 ABILITY TO APPLY THE AXIOMATIC APPROACH IN DEVELOPING PROOFS FOR THEOREMS

SKILL 30.1 State the conditional in the "if-then" form and identify the hypotheses and conclusions.

Conditional statements are frequently written in "if-then" form. The "if" clause of the conditional is known as the **hypothesis**, and the "then" clause is called the **conclusion**. In a proof, the hypothesis is the information that is assumed to be true, while the conclusion is what is to be proven true. A conditional is considered to be of the form:

If p, then q
p is the hypothesis. q is the conclusion.

Conditional statements can be diagrammed using a **Venn diagram**. A diagram can be drawn with one circle inside another circle. The inner circle represents the hypothesis. The outer circle represents the conclusion. If the hypothesis is taken to be true, then you are located inside the inner circle. If you are located in the inner circle then you are also inside the outer circle, so that proves the conclusion is true.

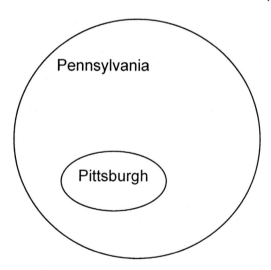

Example:
If an angle has a measure of 90 degrees, then it is a right angle.

> In this statement "an angle has a measure of 90 degrees" is the hypothesis.
> In this statement "it is a right angle" is the conclusion.

MATHEMATICS HIGH SCHOOL

Example:
If you are in Pittsburgh, then you are in Pennsylvania.
  In this statement "you are in Pittsburgh" is the hypothesis.
  In this statement "you are in Pennsylvania" is the conclusion.

SKILL 30.2 Writing the inverse, converse, and contrapositive of a given conditional.

Conditional statements are explained in section 30.1.

Conditional: If p, then q

p is the hypothesis. q is the conclusion.

Inverse: If ~p, then ~q.

Negate both the hypothesis (If not p, then not q) and the conclusion from the original conditional.

Converse : If q, then p.

Reverse the 2 clauses. The original hypothesis becomes the conclusion. The original conclusion then becomes the new hypothesis.

Contrapositive: If ~q, then ~p.

Reverse the 2 clauses. The If not q, then not p original hypothesis becomes the conclusion. The original conclusion then becomes the new hypothesis. THEN negate both the new hypothesis and the new conclusion.

Example: Given the **conditional**:

If an angle has 60°, then it is an acute angle.

Its **inverse**, in the form "If ~p, then ~q", would be:

If an angle doesn't have 60°, then it is not an acute angle.

  NOTICE that the inverse is not true, even
  though the conditional statement was true.

TEACHER CERTIFICATION EXAM

Its **converse**, in the form "If q, then p", would be:

If an angle is an acute angle, then it has 60°.

> NOTICE that the converse is not true, even though the conditional statement was true.

Its **contrapositive**, in the form "If q, then p", would be:

If an angle isn't an acute angle, then it doesn't have 60°.

> NOTICE that the contrapossitive is true, assuming the original conditional statement was true.

TIP: If you are asked to pick a statement that is logically equivalent to a given conditional, look for the contra-positive. The inverse and converse are not always logically equivalent to every conditional. The contra-positive is ALWAYS logically equivalent.

Find the inverse, converse and contrapositive of the following conditional statement. Also determine if each of the 4 statements is true or false.

Conditional: If $x = 5$, then $x^2 - 25 = 0$.   TRUE
Inverse: If $x \neq 5$, then $x^2 - 25 \neq 0$.   FALSE, x could be ⁻5
Converse: If $x^2 - 25 = 0$, then $x = 5$.   FALSE, x could be ⁻5
Contrapositive: If $x^2 - 25 \neq 0$, then $x \neq 5$.   TRUE

Conditional: If $x = 5$, then $6x = 30$.   TRUE
Inverse: If $x \neq 5$, then $6x \neq 30$.   TRUE
Converse: If $6x = 30$, then $x = 5$.   TRUE
Contrapositive: If $6x \neq 30$, then $x \neq 5$.   TRUE

Sometimes, as in this example, all 4 statements can be logically equivalent; however, the only statement that will always be logically equivalent to the original conditional is the contrapositive.

MATHEMATICS HIGH SCHOOL

TEACHER CERTIFICATION EXAM

SKILL 30.3 Draw correct conclusions from given statements.

Conditional statements can be diagrammed using a **Venn diagram**. A diagram can be drawn with one figure inside another figure. The inner figure represents the hypothesis. The outer figure represents the conclusion. If the hypothesis is taken to be true, then you are located inside the inner figure. If you are located in the inner figure then you are also inside the outer figure, so that proves the conclusion is true. Sometimes that conclusion can then be used as the hypothesis for another conditional, which can result in a second conclusion.

Suppose that these statements were given to you, and you are asked to try to reach a conclusion. The statements are:

All swimmers are athletes.
All athletes are scholars.

In "if-then" form, these would be:
If you are a swimmer, then you are an athlete.
If you are an athlete, then you are a scholar.

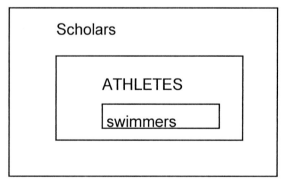

Clearly, if you are a swimmer, then you are also an athlete. This includes you in the group of scholars.

Suppose that these statements were given to you, and you are asked to try to reach a conclusion. The statements are:

All swimmers are athletes.
All wrestlers are athletes.

In "if-then" form, these would be:
If you are a swimmer, then you are an athlete.
If you are a wrestler, then you are an athlete.

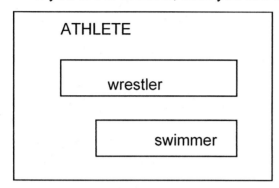

Clearly, if you are a swimmer or a wrestler, then you are also an athlete. This does NOT allow you to come to any other conclusions.

A swimmer may or may NOT also be a wrestler. Therefore, NO CONCLUSION IS POSSIBLE.

Suppose that these statements were given to you, and you are asked to try to reach a conclusion. The statements are:

All rectangles are parallelograms.
Quadrilateral ABCD is not a parallelogram.

In "if-then" form, the first statement would be:
If a figure is a rectangle, then it is also a parallelogram.

Note that the second statement is the negation of the conclusion of statement one. Remember also that the contrapositive is logically equivalent to a given conditional. That is, **"If ~ q, then ~ p"**. Since" ABCD is NOT a parallelogram " is like saying **"If ~ q,"** then you can come to the conclusion **"then ~ p"**. Therefore, the conclusion is ABCD is not a rectangle. Looking at the Venn diagram below, if all rectangles are parallelograms, then rectangles are included as part of the parallelograms. Since quadrilateral ABCD is not a parallelogram, that it is excluded from anywhere inside the parallelogram box. This allows you to conclude that ABCD can not be a rectangle either.

PARALLELOGRAMS
rectangles

quadrilateral ABCD

Try These:

What conclusion, if any, can be reached? Assume each statement is true, regardless of any personal beliefs.

1. If the Red Sox win the World Series, I will die.
   I died.

2. If an angle's measure is between 0° and 90°, then the angle is acute.
   Angle B is not acute.

3. Students who do well in geometry will succeed in college.
   Annie is doing extremely well in geometry.

4. Left-handed people are witty and charming.
   You are left-handed.

SKILL 30.4 Identify point, line, and plane as undefined terms and use synbols for lines, segments, rays, and distances.

The 3 undefined terms of geometry are point, line, and plane.

A plane is a flat surface that extends forever in two dimensions. It has no ends or edges. It has no thickness to it. It is usually drawn as a parallelogram that can be named either by 3 non-collinear points (3 points that are not on the same line) on the plane or by placing a letter in the corner of the plane that is not used elsewhere in the diagram.

A line extends forever in one dimension. It is determined and named by 2 points that are on the line. The line consists of every point that is between those 2 points as well as the points that are on the "straight" extension each way. A line is drawn as a line segment with arrows facing opposite directions on each end to indicate that the line continues in both directions forever.

A point is a position in space, on a line, or on a plane. It has no thickness and no width. Only 1 line can go through any 2 points. A point is represented by a dot named by a single letter.

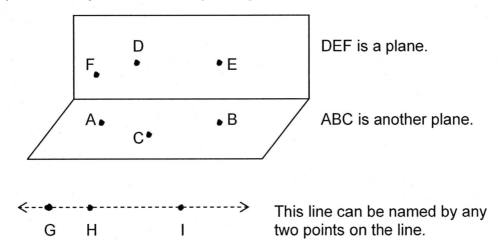

DEF is a plane.

ABC is another plane.

This line can be named by any two points on the line.

It could be named $\overleftrightarrow{GH}$, $\overleftrightarrow{HI}$, $\overleftrightarrow{GI}$, $\overleftrightarrow{IG}$, $\overleftrightarrow{IH}$, or $\overleftrightarrow{HG}$. Any 2 points (letters) on the line can be used and their order is not important in naming a line.

In the above diagrams, A, B, C, D, E, F, G, H, and I are all locations of individual points.

A ray is not an undefined term. A ray consists of all the points on a line starting at one given point and extending in only one of the two opposite directions along the line. The ray is named by naming 2 points on the ray. The first point must be the endpoint of the ray, while the second point can be any other point along the ray. The symbol for a ray is a ray above the 2 letters used to name it. The endpoint of the ray MUST be the first letter.

This ray could be named $\overrightarrow{JK}$ or $\overrightarrow{JL}$. It can not be called $\overrightarrow{KJ}$ or $\overrightarrow{LJ}$ or $\overrightarrow{LK}$ or $\overrightarrow{KL}$ because none of those names start with the endpoint, J.

The **distance** between 2 points on a number line is equal to the absolute value of the difference of the two numbers associated with the points.

If one point is located at "a" and the other point is at "b", then the distance between them is found by this formula:

$$\text{distance} = |a - b| \text{ or } |b - a|$$

If one point is located at $^-3$ and another point is located at 5, the distance between them is found by:

$$\text{distance} = |a - b| = |(^-3) - 5| = |^-8| = 8$$

SKILL 30.5 Differentiate among undefined terms, definitions, postulates, and theorems.

The only undefined terms are point, line and plane.

Definitions are explanations of all mathematical terms except those that are undefined.

Postulates are mathematical statements that are accepted as true statements without providing a proof.

Theorems are mathematical statements that can be proven to be true based on postulates, definitions, algebraic properties, given information, and previously proved theorems.

SKILL 30.6 Write algebraic postulates.

The following algebraic postulates are frequently used as reasons for statements in 2 column geometric properties:

Addition Property:

If $a = b$ and $c = d$, then $a + c = b + d$.

Subtraction Property:

If $a = b$ and $c = d$, then $a - c = b - d$.

Multiplication Property:

If $a = b$ and $c \neq 0$, then $ac = bc$.

Division Property:

If $a = b$ and $c \neq 0$, then $a/c = b/c$.

Reflexive Property: $a = a$
Symmetric Property: If $a = b$, then $b = a$.
Transitive Property: If $a = b$ and $b = c$, then $a = c$.
Distributive Property: $a(b + c) = ab + ac$
Substitution Property: If $a = b$, then b may be substituted for a in any other expression (a may also be substituted for b).

SKILL 30.7 Write simple proofs in two-column form.

In a 2 column proof, the left side of the proof should be the given information, or statements that could be proved by deductive reasoning. The right column of the proof consists of the reasons used to determine that each statement to the left was verifiably true. The right side can identify given information, or state theorems, postulates, definitions or algebraic properties used to prove that particular line of the proof is true.

SKILL 30.8 Write indirect proofs.

Assume the opposite of the conclusion. Keep your hypothesis and given information the same. Proceed to develop the steps of the proof, looking for a statement that contradicts your original assumption or some other known fact. This contradiction indicates that the assumption you made at the beginning of the proof was incorrect; therefore, the original conclusion has to be true.

MATHEMATICS HIGH SCHOOL

SKILL 30.9 Classify conclusions as examples of inductive or deductive thinking.

Inductive thinking is the process of finding a pattern from a group of examples. That pattern is the conclusion that this set of examples seemed to indicate. It may be a correct conclusion or it may be an incorrect conclusion because other examples may not follow the predicted pattern.

Deductive thinking is the process of arriving at a conclusion based on other statements that are all known to be true, such as theorems, axiomspostulates, or postulates. Conclusions found by deductive thinking based on true statements will **always** be true.

Examples :

Suppose:
 On Monday Mr.Peterson eats breakfast at McDonalds.
 On Tuesday Mr.Peterson eats breakfast at McDonalds.
 On Wednesday Mr.Peterson eats breakfast at McDonalds.
 On Thursday Mr.Peterson eats breakfast at McDonalds again.

Conclusion: On Friday Mr. Peterson will eat breakfast at McDonalds again.

This is a conclusion based on inductive reasoning. Based on several days observations, you conclude that Mr. Peterson will eat at McDonalds. This may or may not be true, but it is a conclusion arrived at by inductive thinking.

COMPETENCY 31.0   ABILITY TO APPLY THE PROPERTIES OF LINES, ANGLES, TRIANGLES, QUADRILATERALS, AND CIRCLES IN DEVELOPING APPROPRIATE PROOFS AND SOLVING PROBLEMS.

SKILL 31.1   **Apply SSS, ASA, SAS relationships to proofs for triangles.**

Two triangles are congruent if each of the three angles and three sides of one triangle match up in a one-to-one fashion with congruent angles and sides of the second triangle. In order to see how the sides and angles match up, it is sometimes necessary to imagine rotating or reflecting one of the triangles so the two figures are oriented in the same position.

There are shortcuts to the above procedure for proving two triangles congruent.

Side-Side-Side (SSS) Congruence--If the three sides of one triangle match up in a one-to-one congruent fashion with the three sides of the other triangle, then the two triangles are congruent. With SSS it is not necessary to even compare the angles; they will automatically be congruent.

Angle-Side-Angle (ASA) Congruence--If two angles of one triangle match up in a one-to-one congruent fashion with two angles in the other triangle and if the sides between the two angles are also congruent, then the two triangles are congruent. With ASA the sides that are used for congruence must be located between the two angles used in the first part of the proof.

Side-Angle-Side (SAS) Congruence--If two sides of one triangle match up in a one-to-one congruent fashion with two sides in the other triangle and if the angles between the two sides are also congruent, then the two triangles are congruent. With SAS the angles that are used for congruence must be located between the two sides used in the first part of the proof.

SKILL 31.2 Use the AAS theorem to prove two triangles congruent and HL theorem to prove two right triangles congruent.

In addition to SSS, ASA, and SAS, **Angle-Angle-Side (AAS)** is also a congruence shortcut.

AAS states that if two angles of one triangle match up in a one-to-one congruent fashion with two angle in the other triangle and if two sides that are not between the aforementioned sets of angles are also congruent, then the triangles are congruent. ASA and AAS are very similar; the only difference is where the congruent sides are located. If the sides are between the congruent sets of angles, use ASA. If the sides are not located between the congruent sets of angles, use AAS.

Hypotenuse-Leg (HL) is a congruence shortcut which can only be used with right triangles. If the hypotenuse and leg of one right triangle are congruent to the hypotenuse and leg of the other right triangle, then the two triangles are congruent.

SKILL 31.3 Prove congruence in the case of overlapping triangles.

Two triangles are overlapping if a portion of the interior region of one triangle is shared in common with all or a part of the interior region of the second triangle.

The most effective method for proving two overlapping triangles congruent is to draw the two triangles separated. Separate the two triangles and label all of the vertices using the labels from the original overlapping figures. Once the separation is complete, apply one of the congruence shortcuts: SSS, ASA, SAS, AAS, or HL.

TEACHER CERTIFICATION EXAM

SKILL 31.4 Apply theorems that can be used to prove that a given quadrilateral is a parallelogram.

A parallelogram is a quadrilateral (four-sided figure) in which opposite sides are parallel. There are three shortcuts for proving that a quadrilateral is a parallelgrom without directly showing that the opposite sides are parallel.
If the diagonals of a quadrilateral bisect each other, then the quadrilateral is also a parallelogram. Note that this shortcut only requires the diagonals to bisect each other; the diagonals do not need to be congruent.

If both pairs of opposite sides are congruent, then the quadrilateral is a parallelogram.

If both pairs of opposite angles are congruent, then the quadrilateral is a parallelogram.

If one pair of opposite sides are both parallel and congruent, then the quadrilateral is a parallelogram.

SKILL 31.5 Identify the special properties of parallelograms, rectangles, rhombuses, and squares.

The following table illustrates the properties of each quadrilateral.

	Parallel Opposite Sides	Bisecting Diagonals	Equal Opposite Sides	Equal Opposite Angles	Equal Diagonals	All Sides Equal	All Angles Equal	Perpendicular Diagonals
Parallelogram	X	X	X	X				
Rectangle	X	X	X	X	X		X	
Rhombus	X	X	X	X		X		X
Square	X	X	X	X	X	X	X	X

MATHEMATICS HIGH SCHOOL

TEACHER CERTIFICATION EXAM

SKILL 31.6 Define trapezoids and apply the theorem pertaining to their medians.

A trapezoid is a quadrilateral with exactly one pair of parallel sides. A trapezoid is different from a parallelogram because a parallelogram has two pairs of parallel sides.

The two parallel sides of a trapezoid are called the bases, and the two non-parallel sides are called the legs. If the two legs are the same length, then the trapezoid is called isosceles.
The segment connecting the two midpoints of the legs is called the median. The median has the following two properties.

The median is parallel to the two bases.
The length of the median is equal to one-half the sum of the length of the two bases.

SKILL 31.7 Apply the theorem pertaining to the bisector of an angle of a triangle.

The segment joining the midpoints of two sides of a triangle is called a median. All triangles have three medians. Each median has the following two properties.

A median is parallel to the third side of the triangle.
The length of a median is one-half the length of the third side of the triangle.

SKILL 31.8 Apply the theorems and converses pertaining to a point on the bisector of an angle and a point on the perpendicular bisector of a segment.

Every angle has exactly one ray which bisects the angle. If a point on such a bisector is located, then the point is equidistant from the two sides of the angle. Distance from a point to a side is measured along a segment which is perpendicular to the angle's side. The converse is also true. If a point is equidistant from the sides of an angle, then the point is on the bisector of the angle.

MATHEMATICS HIGH SCHOOL

Every segment has exactly one line which is both perpendicular to and bisects the segment. If a point on such a perpendicular bisector is located, then the point is equidistant to the endpoints of the segment. The converse is also true. If a point is equidistant from the endpoints of a segments, then that point is on the perpendicular bisector of the segment.

SKILL 31.9 Define the median and the altitude of a triangle.

A median is a segment that connects a vertex to the midpoint of the side opposite from that vertex. Every triangle has exactly three medians. An altitude is a segment which extends from one vertex and is perpendicular to the side opposite that vertex. In some cases, the side opposite from the vertex used will need to be extended in order for the altitude to form a perpendicular to the opposite side. The length of the altitude is used when referring to the height of the triangle.

SKILL 31.10 Apply the theorem pertaining to the bisector of an angle of a triangle.

If the three segments which bisect the three angles of a triangle are drawn, the segments will all intersect in a single point. This point is equidistant from all three sides of the triangle. Recall that the distance from a point to a side is measured along the perpendicular from the point to the side.

SKILL 31.11 Apply the theorems pertaining to the intersection of two parallel planes by a third and theorems pertaining to a parallel and a perpendicular to a given line through a point outside the line.

If two planes are parallel and a third plane intersects the first two, then the three planes will intersect in two lines which are also parallel.

Given a line and a point which is not on the line but is in the same plane, then there is exactly one line through the point which is parallel to the given line and exactly one line through the point which is perpendicular to the given line.

SKILL 31.12 Select from the sets of angle bisectors, medians, altitudes, and perpendicular bisectors of sides of a triangle the set(s) that may be or must be concurrent on, in, or outside a triangle.

If three or more segments intersect in a single point, the point is called a point of concurrency.

The following sets of special segments all intersect in points of concurrency.

1. Angle Bisectors
2. Medians
3. Altitudes
4. Perpendicular Bisectors

The points of concurrency can lie inside the triangle, outside the triangle, or on one of the sides of the triangle. The follwing table summarizes this information.

Possible Location(s) of the Points of Concurrency

	Inside the Triangle	Outside the Triangle	On the Triangle
Angle Bisectors	x		
Medians	x		
Altitudes	x	x	x
Perpendicular Bisectors	x	x	x

TEACHER CERTIFICATION EXAM

SKILL 31.13 Identify the center of a circle inscribed in a triangle or circumscribed about a triangle as the point of concurrency of the angle bisectors and perpendicular bisectors of the sides of a triangle.

A circle is inscribed in a triangle if the three sides of the triangle are each tangent to the circle. The center of an inscribed circle is called the incenter of the triangle. To find the incenter, draw the three angle bisectors of the triangle. The point of concurrency of the angle bisectors is the incenter or center of the inscribed circle. Each triangle has only one inscribed circle. A circle is circumscribed about a triangle if the three vertices of the triangle are all located on the circle. The center of a circumscribed circle is called the circumcenter of the triangle. To find the circumcenter, draw the three perpendicular bisectors of the sides of the triangle. The point of concurrency of the perpendicular bisectors is the circumcenter or the center of the circumscribing circle. Each triangle has only one circumscribing circle.

SKILL 31.14 Compute the lengths of the segments of medians as created by the concurrency point of the medians.

A median is a segment which connects a vertex to the midpoint of the side opposite that vertex. Every triangle has three medians. The point of concurrency of the three medians is called the centroid.

The centroid divides each median into two segments whose lengths are always in the ratio of 1:2. The distance from the vertex to the centroid is always twice the distance from the centroid to the midpoint of the side opposite the vertex.

SKILL 31.15 Find the relationships between two circles with given radii.

If two circles have radii which are in a ratio of $a:b$, then the following ratios are also true for the circles.

The diameters are also in the ratio of $a:b$.
The circumferences are also in the ratio $a:b$.
The areas are in the ratio $a^2:b^2$, or the ratio of the areas is the square of the ratios of the radii.

MATHEMATICS HIGH SCHOOL

COMPETENCY 32.0 KNOWLEDGE OF EUCLIDEAN AND NON-EUCLIDEAN GEOMETRIES.

SKILL 32.1 Differentiate between Euclidean and non-Euclidean geometries.

Euclid wrote a set of 13 books around 330 B.C. called the Elements. He outlined ten axioms and then deduced 465 theorems. Euclidean geometry is based on the undefined concept of the point, line and plane.

The fifth of Euclid's axioms (referred to as the parallel postulate) was not as readily accepted as the other nine axioms. Many mathematicians throughout the years have attempted to prove that this axiom is not necessary because it could be proved by the other nine. Among the many who attempted to prove this was Carl Friedrich Gauss. His works led to the development of hyperbolic geometry. Elliptical or Reimannian geometry was suggested by G.F. Berhard Riemann. He based his work on the theory of surfaces and used models as physical interpretations of the undefined terms that satisfy the axioms.

The chart below lists the fifth axiom (parallel postulate) as it is given in each of the three geometries.

EUCLIDEAN	ELLIPTICAL	HYPERBOLIC
Given a line and a point not on that line, one and only one line can be drawn through the given point parallel to the given line.	Given a line and a point not on that line, no line can be drawn through the given point parallel to the given line.	Given a line and a point not on that line, two or more lines can be drawn through the point parallel to the given line.

TEACHER CERTIFICATION EXAM

COMPETENCY 33.0 **DETERMINE PERIMETER AND AREA OF PLANE FIGURES AND SURFACE AREA AND VOLUME OF REGULAR SOLID FIGURES.**

SKILL 33.1 Identify convex polygons and a regular polygons.

In order to determine if a figure is convex and then determine if it is regular, it is necessary to apply the definition of convex first.

Convex polygons: polygons in which no line containing the side of the polygon contains a point on the interior of the polygon.

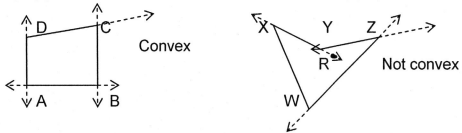

Regular polygons: convex polygons in which all sides are congruent and all angles are congruent (in other words, a regular polygon must be both equilateral and equiangular).

SKILL 33.2 Demonstrate the ability to find surface area and volume of regular solid figures.

Use the formulas to find the volume and surface area.

FIGURE	VOLUME	TOTAL SURFACE AREA
Right Cylinder	$\pi r^2 h$	$2\pi rh + 2\pi r^2$
Right Cone	$\dfrac{\pi r^2 h}{3}$	$\pi r \sqrt{r^2 + h^2} + \pi r^2$
Sphere	$\dfrac{4}{3}\pi r^3$	$4\pi r^2$
Rectangular Solid	LWH	$2LW + 2WH + 2LH$

Note: $\sqrt{r^2 + h^2}$ is equal to the slant height of the cone.

Sample problem:

1. Given the figure below, find the volume and surface area.

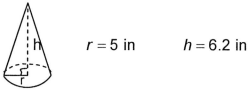

$r = 5$ in $h = 6.2$ in

Volume $= \dfrac{\pi r^2 h}{3}$ First write the formula.

$\dfrac{1}{3}\pi(5^2)(6.2)$ Then substitute.

162.31562 cubic inches   Finally solve the problem.

Surface area $= \pi r\sqrt{r^2 + h^2} + \pi r^2$ First write the formula.

$\pi 5\sqrt{5^2 + 6.2^2} + \pi 5^2$ Then substitute.

203.652 square inches   Compute.

Note: volume is always given in cubic units and area is always given in square units.

SKILL 33.3 Apply area formulas and determine the perimeter of a polygon.

FIGURE	AREA FORMULA	PERIMETER FORMULA
Rectangle	LW	$2(L+W)$
Triangle	$\dfrac{1}{2}bh$	$a+b+c$
Parallelogram	bh	sum of lengths of sides
Trapezoid	$\dfrac{1}{2}h(a+b)$	sum of lengths of sides

MATHEMATICS HIGH SCHOOL

Sample problems:

1. Find the area and perimeter of a rectangle if its length is 12 inches and its diagonal is 15 inches.

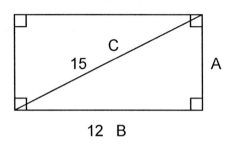

1. Draw and label sketch.

2. Since the height is still needed use Pythagorean formula to find missing leg of the triangle.

$$A^2 + B^2 = C^2$$
$$A^2 + 12^2 = 15^2$$
$$A^2 = 15^2 - 12^2$$
$$A^2 = 81$$
$$A = 9$$

Now use this information to find the area and perimeter.

$A = LW$	$P = 2(L + W)$	1. write formula
$A = (12)(9)$	$P = 2(12 + 9)$	2. substitute
$A = 108 \text{ in}^2$	$P = 42$ inches	3. solve

SKILL 33.4 Apply the formulas for determining the circumferences and areas of circles.

Given a circular figure the formulas are as follows:

$$A = \pi r^2 \qquad C = \pi d \text{ or } 2\pi r$$

Sample problem:

1. If the area of a circle is 50 cm², find the circumference.

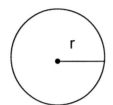　　A = 50 cm²

1. Draw sketch.
2. Determine what is still needed.

Use the area formula to find the radius.

$A = \pi r^2$	1. write formula
$50 = \pi r^2$	2. substitute
$\dfrac{50}{\pi} = r^2$	3. divide by π
$15.915 = r^2$	4. substitute
$\sqrt{15.915} = \sqrt{r^2}$	5. take square root of both sides
$3.989 \approx r$	6. compute

Use the approximate answer (due to rounding) to find the circumference.

$C = 2\pi r$	1. write formula
$C = 2\pi (3.989)$	2. substitute
$C \approx 25.064$	3. compute

TEACHER CERTIFICATION EXAM

SKILL 33.5 **Find the areas of parallelograms, triangles, and trapezoids using area formulas.**

When using formulas to find each of the required items it is helpful to remember to always use the same strategies for problem solving. First, draw and label a sketch if needed. Second, write the formula down and then substitute in the known values. This will assist in identifying what is still needed (the unknown). Finally, solve the resulting equation.

Being consistent in the strategic approach to problem solving is paramount to teaching the concept as well as solving it.

SKILL 33.6 **Apply the area formulas to figures composed of parallelograms, triangles and trapezoids.**

Use appropriate problem solving strategies to find the solution.

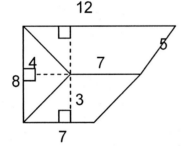

1. Find the area of the given figure.

2. Cut the figure into familiar shapes.

3. Identify what type figures are given and write the appropriate formulas.

Area of figure 1 (triangle)	Area of figure 2 (parallelogram)	Area of figure 3 (trapezoid)
$A = \frac{1}{2}bh$	$A = bh$	$A = \frac{1}{2}h(a+b)$
$A = \frac{1}{2}(8)(4)$	$A = (7)(3)$	$A = \frac{1}{2}(5)(12+7)$
$A = 16$ sq. ft	$A = 21$ sq. ft	$A = 47.5$ sq. ft

Now find the total area by adding the area of all figures.

Total area = 16 + 21 + 47.5
Total area = 84.5 square ft

MATHEMATICS HIGH SCHOOL

TEACHER CERTIFICATION EXAM

SKILL 33.7 Identify and use the parts of a regular polygon to compute the area of a polygon.

Given the figure below, find the area by dividing the polygon into smaller shapes.

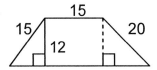

1. divide the figure into two triangles and a rectangle.

2. find the missing lengths.

3. find the area of each part.

4. find the sum of all areas.

Find base of both right triangles using Pythagorean Formula:

$$a^2 + b^2 = c^2 \qquad\qquad a^2 + b^2 = c^2$$
$$a^2 + 12^2 = 15^2 \qquad a^2 + 12^2 = 20^2$$
$$a^2 = 225 - 144 \qquad a^2 = 400 - 144$$
$$a^2 = 81 \qquad\qquad a^2 = 256$$
$$a = 9 \qquad\qquad\quad a = 16$$

Area of triangle 1   Area of triangle 2   Area of rectangle

$$A = \tfrac{1}{2}bh \qquad\qquad A = \tfrac{1}{2}bh \qquad\qquad A = LW$$
$$A = \tfrac{1}{2}(9)(12) \qquad A = \tfrac{1}{2}(16)(12) \qquad A = (15)(12)$$
$$A = 54 \text{ sq. units} \quad A = 96 \text{ sq. units} \quad A = 180 \text{ sq. units}$$

Find the sum of all three figures.

$$54 + 96 + 180 = 330 \text{ square units}$$

SKILL 33.8 Compare the perimeters and areas of similar polygons.

Polygons are similar if and only if there is a one-to-one correspondence between their vertices such that the corresponding angles are congruent and the lengths of corresponding sides are proportional.

Given the rectangles below, compare the area and perimeter.

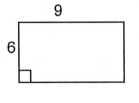
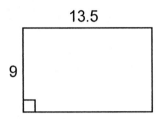

$A = LW$	$A = LW$	1. write formula
$A = (6)(9)$	$A = (9)(13.5)$	2. substitute known values
$A = 54$ sq. units	$A = 121.5$ sq. units	3. compute
$P = 2(L + W)$	$P = 2(L + W)$	1. write formula
$P = 2(6 + 9)$	$P = 2(9 + 13.5)$	2. substitute known values
$P = 30$ units	$P = 45$ units	3. compute

Notice that the areas relate to each other in the following manner:

Ratio of sides $\quad 9/13.5 = 2/3$

Multiply the first area by the square of the reciprocal $(3/2)^2$ to get the second area.
$$54 \times (3/2)^2 = 121.5$$

The perimeters relate to each other in the following manner:

Ratio of sides $\quad 9/13.5 = 2/3$

Multiply the perimeter of the first by the reciprocal of the ratio to get the perimeter of the second.

$$30 \times 3/2 = 45$$

SKILL 33.9 Apply the formulas for lateral area, total area and volume to right prisms and regular pyramids.

FIGURE	LATERAL AREA	TOTAL AREA	VOLUME
Right prism	sum of area of lateral faces (rectangles)	lateral area plus 2 times the area of base	area of base times height
regular pyramid	sum of area of lateral faces (triangles)	lateral area plus area of base	1/3 times the area of the base times the height

Find the total area of the given figure:

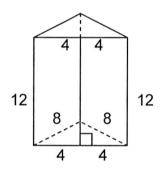

1. Since this is a triangular prism, first find the area of the bases.

2. Find the area of each rectangular lateral face.

3. Add the areas together.

$A = \dfrac{1}{2}bh$ $A = LW$ 1. write formula

$8^2 = 4^2 + h^2$ 2. find the height of
$h = 6.928$ the base triangle

$A = \dfrac{1}{2}(8)(6.928)$ $A = (8)(12)$ 3. substitute known values

$A = 27.713$ sq. units $A = 96$ sq. units   4. compute

Total Area $= 2(27.713) + 3(96)$
$= 343.426$ sq. units

TEACHER CERTIFICATION EXAM

SKILL 33.10 Apply the formulas for lateral area, total area and volume to right circular cylinders and cones.

FIGURE	VOLUME	TOTAL SURFACE AREA	LATERAL AREA
Right Cylinder	$\pi r^2 h$	$2\pi rh + 2\pi r^2$	$2\pi rh$
Right Cone	$\dfrac{\pi r^2 h}{3}$	$\pi r\sqrt{r^2 + h^2} + \pi r^2$	$\pi r\sqrt{r^2 + h^2}$

Note: $\pi r\sqrt{r^2 + h^2}$ is equal to the slant height of the cone.

Sample problem:

1. A water company is trying to decide whether to use traditional cylindrical paper cups or to offer conical paper cups since both cost the same. The traditional cups are 8 cm wide and 14 cm high. The conical cups are 12 cm wide and 19 cm high. The company will use the cup that holds the most water.

1. Draw and label a sketch of each.

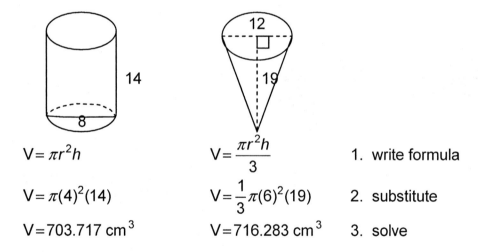

$V = \pi r^2 h$	$V = \dfrac{\pi r^2 h}{3}$	1. write formula
$V = \pi(4)^2(14)$	$V = \dfrac{1}{3}\pi(6)^2(19)$	2. substitute
$V = 703.717 \text{ cm}^3$	$V = 716.283 \text{ cm}^3$	3. solve

The choice should be the conical cup since its volume is more.

MATHEMATICS HIGH SCHOOL

TEACHER CERTIFICATION EXAM

SKILL 33.11 Apply the formulas for surface area and volume of spheres.

FIGURE	VOLUME	TOTAL SURFACE AREA
Sphere	$\frac{4}{3}\pi r^3$	$4\pi r^2$

Sample problem:

1. How much material is needed to make a basketball that has a diameter of 15 inches? How much air is needed to fill the basketball?

Draw and label a sketch:

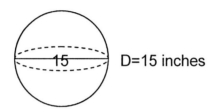 D=15 inches

Total surface area   Volume

$TSA = 4\pi r^2$ $V = \frac{4}{3}\pi r^3$ 1. write formula

$= 4\pi(7.5)^2$ $= \frac{4}{3}\pi(7.5)^3$ 2. substitute

$= 706.858 \text{ in}^2$ $= 1767.1459 \text{ in}^3$ 3. solve

MATHEMATICS HIGH SCHOOL

TEACHER CERTIFICATION EXAM

Competency 34.0 ABILITY TO COMPUTE DECRIPTIVE MEASURES FROM GIVEN UNGROUPED DATA.

SKILL 34.1 Find the ranges, variances, and standard deviations for ungrouped data.

An understanding of the definitions is important in determining the validity and uses of statistical data. All definitions and applications in this section apply to ungrouped data.

Data item: each piece of data is represented by the letter X.

Mean: the average of all data represented by the symbol $\overline{X}$.

Range: difference between the highest and lowest value of data items.

Sum of the Squares: sum of the squares of the differences between each item and the mean. $Sx^2 = (X - \overline{X})^2$

Variance: the sum of the squares quantity divided by the number of items. (the lower case greek letter sigma squared (σ^2) represents variance). $\dfrac{Sx^2}{N} = \sigma^2$

The larger the value of the variance the larger the spread

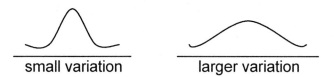

small variation   larger variation

Standard Deviation: the square root of the variance. The lower case greek letter sigma (σ) is used to represent standard deviation. $\sigma = \sqrt{\sigma^2}$

Most statistical calculators have standard deviation keys on them and should be used when asked to calculate statistical functions. It is important to become familiar with the calculator and the location of the keys needed.

MATHEMATICS HIGH SCHOOL

Sample Problem:

Given the ungrouped data below, calculate the mean, range, standard deviation and the variance.

| 15 | 22 | 28 | 25 | 34 | 38 |
| 18 | 25 | 30 | 33 | 19 | 23 |

Mean ($\overline{X}$) = 25.8333333
Range: $38 - 15 = 23$
standard deviation (σ) = 6.6936952
Variance (σ^2) = 44.805556

TEACHER CERTIFICATION EXAM

COMPETENCY 35.0   UNDERSTANDING OF MATRIX ALGEBRA.

SKILL 35.1 Find the value of the determinant of a two-by-two or three-by-three matrix.

A matrix is a square array of numbers called its entries or elements. The dimensions of a matrix are written as the number of rows (r) by the number of columns (r × c).

$$\begin{pmatrix} 1 & 2 & 3 \\ 4 & 5 & 6 \end{pmatrix}$$ is a 2 × 3 matrix (2 rows by 3 columns)

$$\begin{pmatrix} 1 & 2 \\ 3 & 4 \\ 5 & 6 \end{pmatrix}$$ is a 3 × 2 matrix (3 rows by 2 columns)

Associated with every square matrix is a number called the determinant. Use these formulas to calculate determinants.

2 × 2 $\begin{pmatrix} a & b \\ c & d \end{pmatrix} = ad - bc$

3 × 3 $\begin{pmatrix} a_1 & b_1 & c_1 \\ a_2 & b_2 & c_2 \\ a_3 & b_3 & c_3 \end{pmatrix} = (a_1 b_2 c_3 + b_1 c_2 a_3 + c_1 a_2 b_3) - (a_3 b_2 c_1 + b_3 c_2 a_1 + c_3 a_2 b_1)$

This is found by repeating the first two columns and then using the diagonal lines to find the value of each expression as shown below:

$\begin{pmatrix} a_1^* & b_1^\circ & c_1^\bullet \\ a_2 & b_2^* & c_2^\circ \\ a_3 & b_3 & c_3^* \end{pmatrix} \begin{matrix} a_1 & b_1 \\ a_2^\bullet & b_2 \\ a_3^\circ & b_3^\bullet \end{matrix} = (a_1 b_2 c_3 + b_1 c_2 a_3 + c_1 a_2 b_3) - (a_3 b_2 c_1 + b_3 c_2 a_1 + c_3 a_2 b_1)$

Sample Problem:
1. Find the value of the determinant:

$\begin{pmatrix} 4 & ^-8 \\ 7 & 3 \end{pmatrix} = (4)(3) - (7)(^-8)$ Cross multiply and subtract.

$12 - (^-56) = 68$ Then simplify.

MATHEMATICS HIGH SCHOOL

SKILL 35.2 Find the sums and differences of matrices.

Addition of matrices is accomplished by adding the corresponding elements of the two matrices. Subtraction is defined as the inverse of addition. In other words, change the sign on all the elements in the second matrix and add the two matrices.

Sample problems:
Find the sum or difference.

1. $\begin{pmatrix} 2 & 3 \\ {}^-4 & 7 \\ 8 & {}^-1 \end{pmatrix} + \begin{pmatrix} 8 & {}^-1 \\ 2 & {}^-1 \\ 3 & {}^-2 \end{pmatrix} =$

$\begin{pmatrix} 2+8 & 3+({}^-1) \\ {}^-4+2 & 7+({}^-1) \\ 8+3 & {}^-1+({}^-2) \end{pmatrix}$ Add corresponding elements.

$\begin{pmatrix} 10 & 2 \\ {}^-2 & 6 \\ 11 & {}^-3 \end{pmatrix}$ Simplify.

2. $\begin{pmatrix} 8 & {}^-1 \\ 7 & 4 \end{pmatrix} - \begin{pmatrix} 3 & 6 \\ {}^-5 & 1 \end{pmatrix} =$

$\begin{pmatrix} 8 & {}^-1 \\ 7 & 4 \end{pmatrix} + \begin{pmatrix} {}^-3 & {}^-6 \\ 5 & {}^-1 \end{pmatrix} =$ Change all of the signs in the second matrix and then add the two matrices.

$\begin{pmatrix} 8+({}^-3) & {}^-1+({}^-6) \\ 7+5 & 4+({}^-1) \end{pmatrix} =$ Simplify.

$\begin{pmatrix} 5 & {}^-7 \\ 12 & 3 \end{pmatrix}$

MATHEMATICS HIGH SCHOOL

TEACHER CERTIFICATION EXAM

Practice problems:

1. $\begin{pmatrix} 8 & ^-1 \\ 5 & 3 \end{pmatrix} + \begin{pmatrix} 3 & 8 \\ 6 & ^-2 \end{pmatrix} =$

2. $\begin{pmatrix} 3 & 7 \\ ^-4 & 12 \\ 0 & ^-5 \end{pmatrix} - \begin{pmatrix} 3 & 4 \\ 6 & ^-1 \\ ^-5 & ^-5 \end{pmatrix} =$

SKILL 35.3 Find the product of a scalar and a matrix.

Scalar multiplication is the product of the scalar (the outside number) and each element inside the matrix.

Sample problem:

Given: $A = \begin{pmatrix} 4 & 0 \\ 3 & ^-1 \end{pmatrix}$ Find 2A.

$2A = 2 \begin{pmatrix} 4 & 0 \\ 3 & ^-1 \end{pmatrix}$

$\begin{pmatrix} 2 \times 4 & 2 \times 0 \\ 2 \times 3 & 2 \times ^-1 \end{pmatrix}$ Multiply each element in the matrix by the scalar.

$\begin{pmatrix} 8 & 0 \\ 6 & ^-2 \end{pmatrix}$ Simplify.

MATHEMATICS HIGH SCHOOL

Practice problems:

1. $-2\begin{pmatrix} 2 & 0 & 1 \\ -1 & -2 & 4 \end{pmatrix}$

2. $3\begin{pmatrix} 6 \\ 2 \\ 8 \end{pmatrix} + 4\begin{pmatrix} 0 \\ 7 \\ 2 \end{pmatrix}$

3. $2\begin{pmatrix} -6 & 8 \\ -2 & -1 \\ 0 & 3 \end{pmatrix}$

SKILL 35.4 Solve simple matrix equations.

The variable in a matrix equation represents a matrix. When solving for the answer use the adding, subtracting and scalar multiplication properties.

Sample problem:

Solve the matrix equation for the variable X.

$2x + \begin{pmatrix} 4 & 8 & 2 \\ 7 & 3 & 4 \end{pmatrix} = 2\begin{pmatrix} 1 & -2 & 0 \\ 3 & -5 & 7 \end{pmatrix}$

$2x = 2\begin{pmatrix} 1 & -2 & 0 \\ 3 & -5 & 7 \end{pmatrix} - \begin{pmatrix} 4 & 8 & 2 \\ 7 & 3 & 4 \end{pmatrix}$ Subtract $\begin{pmatrix} 4 & 8 & 2 \\ 7 & 3 & 4 \end{pmatrix}$ from both sides.

$2x = \begin{pmatrix} 2 & -4 & 0 \\ 6 & -10 & 14 \end{pmatrix} + \begin{pmatrix} -4 & -8 & -2 \\ -7 & -3 & -4 \end{pmatrix}$ Scalar multiplication and matrix subtraction.

$2x = \begin{pmatrix} -2 & -12 & -2 \\ -1 & -13 & 10 \end{pmatrix}$ Matrix addition.

$x = \begin{pmatrix} -1 & -6 & -1 \\ -\dfrac{1}{2} & -\dfrac{13}{2} & 5 \end{pmatrix}$ Multiply both sides by $\dfrac{1}{2}$.

Solve for the unknown values of the elements in the matrix.

$$\begin{pmatrix} x+3 & y-2 \\ z+3 & w-4 \end{pmatrix} + \begin{pmatrix} -2 & 4 \\ 2 & 5 \end{pmatrix} = \begin{pmatrix} 4 & 8 \\ 6 & 1 \end{pmatrix}$$

$$\begin{pmatrix} x+1 & y+2 \\ z+5 & w+1 \end{pmatrix} = \begin{pmatrix} 4 & 8 \\ 6 & 1 \end{pmatrix} \qquad \text{Matrix addition.}$$

$$\begin{array}{llll} x+1=4 & y+2=8 & z+5=6 & w+1=1 \\ x=3 & y=6 & z=1 & w=0 \end{array} \quad \text{Definition of equal matrices.}$$

Practice problems:

1. $x + \begin{pmatrix} 7 & 8 \\ 3 & ^-1 \\ 2 & ^-3 \end{pmatrix} = \begin{pmatrix} 0 & 8 \\ ^-9 & ^-4 \\ 8 & 2 \end{pmatrix}$

2. $4x - 2\begin{pmatrix} 0 & 10 \\ 6 & ^-4 \end{pmatrix} = 3\begin{pmatrix} 4 & 9 \\ 0 & 12 \end{pmatrix}$

3. $\begin{pmatrix} 7 & 3 \\ 2 & 4 \\ 3 & 7 \end{pmatrix} + \begin{pmatrix} a+2 & b+4 \\ c-3 & d+1 \\ e & f+3 \end{pmatrix} = \begin{pmatrix} 4 & 6 \\ ^-1 & 1 \\ 3 & 0 \end{pmatrix}$

TEACHER CERTIFICATION EXAM

SKILL 35.5 **Find the product of two matrices.**

The product of two matrices can only be found if the number of columns in the first matrix is equal to the number of rows in the second matrix. Matrix multiplication is not necessarily commutative.

Sample problems:

1. Find the product AB if:

$$A = \begin{pmatrix} 2 & 3 & 0 \\ 1 & -4 & -2 \\ 0 & 1 & 1 \end{pmatrix} \qquad B = \begin{pmatrix} -2 & 3 \\ 6 & -1 \\ 0 & 2 \end{pmatrix}$$

$\qquad\qquad 3 \times 3 \qquad\qquad\qquad\qquad 3 \times 2$

Note: Since the number of columns in the first matrix (3 × <u>3</u>) matches the number of rows (<u>3</u> ×2) this product is defined and can be found. The dimensions of the product will be equal to the number of rows in the first matrix (<u>3</u> ×3) by the number of columns in the second matrix (3 × <u>2</u>). The answer will be a 3 × 2 matrix.

$$AB = \begin{pmatrix} 2 & 3 & 0 \\ 1 & -4 & -2 \\ 0 & 1 & 1 \end{pmatrix} \times \begin{pmatrix} -2 & 3 \\ 6 & -1 \\ 0 & 2 \end{pmatrix}$$

$\begin{pmatrix} -2(-2)+3(6)+0(0) & \\ & \\ & \end{pmatrix}$ Multiply 1st row of A by 1st column of B.

$\begin{pmatrix} 14 & 2(3)+3(-1)+0(2) \\ & \\ & \end{pmatrix}$ Multiply 1st row of A by 2nd column of B.

$\begin{pmatrix} 14 & 3 \\ 1(-2)-4(6)-2(0) & \\ & \end{pmatrix}$ Multiply 2nd row of A by 1st column of B.

MATHEMATICS HIGH SCHOOL

$$\begin{pmatrix} 14 & 3 \\ ^-26 & 1(3)-4(^-1)-2(2) \end{pmatrix}$$ Multiply 2nd row of A by 2nd column of B.

$$\begin{pmatrix} 14 & 3 \\ ^-26 & 3 \\ 0(^-2)+1(6)+1(0) & \end{pmatrix}$$ Multiply 3rd row of A by 1st column of B.

$$\begin{pmatrix} 14 & 3 \\ ^-26 & 3 \\ 6 & 0(3)+1(^-1)+1(2) \end{pmatrix}$$ Multiply 3rd row of A by 2nd column of B.

$$\begin{pmatrix} 14 & 3 \\ ^-26 & 3 \\ 6 & 1 \end{pmatrix}$$

The product of BA is not defined since the number of columns in B is not equal to the number of rows in A.

Practice problems:

1. $\begin{pmatrix} 3 & 4 \\ ^-2 & 1 \end{pmatrix} \begin{pmatrix} ^-1 & 7 \\ ^-3 & 1 \end{pmatrix}$

2. $\begin{pmatrix} 1 & ^-2 \\ 3 & 4 \\ 2 & 5 \\ -1 & 6 \end{pmatrix} \begin{pmatrix} 3 & ^-1 & ^-4 \\ ^-1 & 2 & 3 \end{pmatrix}$

SKILL 35.6 Convert a matrix equation to an equivalent system of linear equations or vice versa.

When given the following system of equations:

$$ax + by = e$$
$$cx + dy = f$$

the matrix equation is written in the form:

$$\begin{pmatrix} a & b \\ c & d \end{pmatrix} \begin{pmatrix} x \\ y \end{pmatrix} = \begin{pmatrix} e \\ f \end{pmatrix}$$

The solution is found using the inverse of the matrix of coefficients. Inverse of matrices can be written as follows:

$$A^{-1} = \frac{1}{\text{determinant of } A} \begin{pmatrix} d & -b \\ -c & a \end{pmatrix}$$

Sample Problem:
1. Write the matrix equation of the system.

$$3x - 4y = 2$$
$$2x + y = 5$$

$$\begin{pmatrix} 3 & -4 \\ 2 & 1 \end{pmatrix} \begin{pmatrix} x \\ y \end{pmatrix} = \begin{pmatrix} 2 \\ 5 \end{pmatrix}$$ Definition of matrix equation.

$$\begin{pmatrix} x \\ y \end{pmatrix} = \frac{1}{11} \begin{pmatrix} 1 & 4 \\ -2 & 3 \end{pmatrix} \begin{pmatrix} 2 \\ 5 \end{pmatrix}$$ Multiply by the inverse of the coefficient matrix.

$$\begin{pmatrix} x \\ y \end{pmatrix} = \frac{1}{11} \begin{pmatrix} 22 \\ 11 \end{pmatrix}$$ Matrix multiplication.

$$\begin{pmatrix} x \\ y \end{pmatrix} = \begin{pmatrix} 2 \\ 1 \end{pmatrix}$$ Scalar multiplication.

The solution is (2,1).

Practice problems:

1. $x + 2y = 5$
   $3x + 5y = 14$

2. $-3x + 4y - z = 3$
   $x + 2y - 3z = 9$
   $y - 5z = -1$

TEACHER CERTIFICATION EXAM

COMPETENCY 36.0   **COMPREHENSION OF CIRCULAR/TRIGONOMETRIC FUNCTIONS AND THEIR INVERSES.**

SKILL 36.1 Solve equations involving circular/trigonometric functions and their inverses.

Unlike trigonometric identities that are true for all values of the defined variable, trigonometric equations are true for some, but not all, of the values of the variable. Most often trigonometric equations are solved for values between 0 and 360 degrees or 0 and 2π radians.

Some algebraic operation, such as squaring both sides of an equation, will give you extraneous answers. You must remember to check all solutions to be sure that they work.

Sample problems:

1. Solve: $\cos x = 1 - \sin x$ if $0 \leq x < 360$ degrees.

$\cos^2 x = (1-\sin x)^2$	1. square both sides
$1 - \sin^2 x = 1 - 2\sin x + \sin^2 x$	2. substitute
$0 = {}^-2\sin x + 2\sin^2 x$	3. set = to 0
$0 = 2\sin x({}^-1 + \sin x)$	4. factor
$2\sin x = 0 \quad\quad {}^-1 + \sin x = 0$	5. set each factor = 0
$\sin x = 0 \quad\quad\quad \sin x = 1$	6. solve for $\sin x$
$x = 0$ or $180 \quad\quad x = 90$	7. find value of sin at x

The solutions appear to be 0, 90 and 180. Remember to check each solution and you will find that 180 does not give you a true equation. Therefore, the only solutions are 0 and 90 degrees.

2. Solve: $\cos^2 x = \sin^2 x$ if $0 \leq x < 2\pi$

$\cos^2 x = 1 - \cos^2 x$	1. substitute
$2\cos^2 x = 1$	2. simplify
$\cos^2 x = \dfrac{1}{2}$	3. divide by 2
$\sqrt{\cos^2 x} = \pm\sqrt{\dfrac{1}{2}}$	4. take square root
$\cos x = \dfrac{\pm\sqrt{2}}{2}$	5. rationalize denominator
$x = \dfrac{\pi}{4}, \dfrac{3\pi}{4}, \dfrac{5\pi}{4}, \dfrac{7\pi}{4}$	

MATHEMATICS HIGH SCHOOL

TEACHER CERTIFICATION EXAM

COMPETENCY 37.0   **ABILITY TO PROVE CURCULAR/TRIGONOMETRIC FUNCTION IDENTITIES.**

SKILL 37.1   Prove circular/trigonometric function identities.

Given 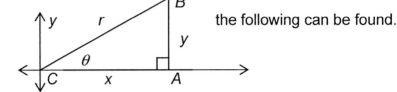 the following can be found.

Trigonometric Functions:

$\sin\theta = \dfrac{y}{r}$ $\csc\theta = \dfrac{r}{y}$

$\cos\theta = \dfrac{x}{r}$ $\sec\theta = \dfrac{r}{x}$

$\tan\theta = \dfrac{y}{x}$ $\cot\theta = \dfrac{x}{y}$

Sample problem:

1. Prove that $\sec\theta = \dfrac{1}{\cos\theta}$.

$\sec\theta = \dfrac{1}{\frac{x}{r}}$ Substitution definition of cosine.

$\sec\theta = \dfrac{1 \times r}{\frac{x}{r} \times r}$ Multiply by $\dfrac{r}{r}$.

$\sec\theta = \dfrac{r}{x}$ Substitution.

$\sec\theta = \sec\theta$ Substitute definition of $\dfrac{r}{x}$.

$\sec\theta = \dfrac{1}{\cos\theta}$ Subsitute.

MATHEMATICS HIGH SCHOOL

2. Prove that $\sin^2 + \cos^2 = 1$.

$\left(\dfrac{y}{r}\right)^2 + \left(\dfrac{x}{r}\right)^2 = 1$ Substitute definitions of sin and cos.

$\dfrac{y^2 + x^2}{r^2} = 1$ $x^2 + y^2 = r^2$ Pythagorean formula.

$\dfrac{r^2}{r^2} = 1$ Simplify.

$1 = 1$ Substitute.

$\sin^2 \theta + \cos^2 \theta = 1$

Practice problems: Prove each identity.

1. $\cot \theta = \dfrac{\cos \theta}{\sin \theta}$ 2. $1 + \cot^2 \theta + \csc^2 \theta$

SKILL 37.2 Apply basic circular/trigonometric function identities.

There are two methods that may be used to prove trigonometric identities. One method is to choose one side of the equation and manipulate it until it equals the other side. The other method is to replace expressions on both sides of the equation with equivalent expressions until both sides are equal.

The Reciprocal Identities

$\sin x = \dfrac{1}{\csc x}$ $\sin x \csc x = 1$ $\csc x = \dfrac{1}{\sin x}$

$\cos x = \dfrac{1}{\sec x}$ $\cos x \sec x = 1$ $\sec x = \dfrac{1}{\cos x}$

$\tan x = \dfrac{1}{\cot x}$ $\tan x \cot x = 1$ $\cot x = \dfrac{1}{\tan x}$

$\tan x = \dfrac{\sin x}{\cos x}$ $\cot x = \dfrac{\cos x}{\sin x}$

The Pythagorean Identities

$\sin^2 x + \cos^2 x = 1$ $1 + \tan^2 x = \sec^2 x$ $1 + \cot^2 x = \csc^2 x$

Sample problems:

1. Prove that $\cot x + \tan x = (\csc x)(\sec x)$.

$\dfrac{\cos x}{\sin x} + \dfrac{\sin x}{\cos x}$ Reciprocal identities.

$\dfrac{\cos^2 x + \sin^2 x}{\sin x \cos x}$ Common denominator.

$\dfrac{1}{\sin x \cos x}$ Pythagorean identity.

$\dfrac{1}{\sin x} \times \dfrac{1}{\cos x}$

$\csc x (\sec x) = \csc x (\sec x)$ Reciprocal identity, therefore,

$\cot x + \tan x = \csc x (\sec x)$

2. Prove that $\dfrac{\cos^2 \theta}{1 + 2\sin\theta + \sin^2 \theta} = \dfrac{\sec\theta - \tan\theta}{\sec\theta + \tan\theta}$.

$\dfrac{1 - \sin^2 \theta}{(1 + \sin\theta)(1 + \sin\theta)} = \dfrac{\sec\theta - \tan\theta}{\sec\theta + \tan\theta}$ Pythagorean identity factor denominator.

$\dfrac{1 - \sin^2 \theta}{(1 + \sin\theta)(1 + \sin\theta)} = \dfrac{\dfrac{1}{\cos\theta} - \dfrac{\sin\theta}{\cos\theta}}{\dfrac{1}{\cos\theta} + \dfrac{\sin\theta}{\cos\theta}}$ Reciprocal identities.

$\dfrac{(1 - \sin\theta)(1 + \sin\theta)}{(1 + \sin\theta)(1 + \sin\theta)} = \dfrac{\dfrac{1 - \sin\theta}{\cos\theta}(\cos\theta)}{\dfrac{1 + \sin\theta}{\cos\theta}(\cos\theta)}$ Factor $1 - \sin^2\theta$.

Multiply by $\dfrac{\cos\theta}{\cos\theta}$.

$\dfrac{1 - \sin\theta}{1 + \sin\theta} = \dfrac{1 - \sin\theta}{1 + \sin\theta}$ Simplify.

$\dfrac{\cos^2 \theta}{1 + 2\sin\theta + \sin^2 \theta} = \dfrac{\sec\theta - \tan\theta}{\sec\theta + \tan\theta}$

COMPETENCY 38.0　　PROFICIENCY IN GRAPHING TRIGONOMETRIC FUNCTIONS.

SKILL 38.1　Graph trigonometric functions.

It is easiest to graph trigonometric functions when using a calculator by making a table of values.

DEGREES

	0	30	45	60	90	120	135	150	180	210	225	240	270	300	315	330	360
sin	0	.5	.71	.87	1	.87	.71	.5	0	-.5	-.71	-.87	-1	-.87	-.71	-.5	0
cos	1	.87	.71	.5	0	-.5	-.71	-.87	-1	-.87	-.71	-.5	0	.5	.71	.87	1
tan	0	.57	1	1.7	--	-1.7	-1	-.57	0	.57	1	1.7	--	-1.7	-1	-.57	0
	0	$\frac{\pi}{6}$	$\frac{\pi}{4}$	$\frac{\pi}{3}$	$\frac{\pi}{2}$	$\frac{2\pi}{3}$	$\frac{3\pi}{4}$	$\frac{5\pi}{6}$	π	$\frac{7\pi}{6}$	$\frac{5\pi}{4}$	$\frac{4\pi}{3}$	$\frac{3\pi}{2}$	$\frac{5\pi}{3}$	$\frac{7\pi}{4}$	$\frac{11\pi}{6}$	2π

RADIANS

Remember the graph always ranges from +1 to ⁻1 for sine and cosine functions unless noted as the coefficient of the function in the equation. For example, $y = 3\cos x$ has an amplitude of 3 units from the center line (0).

Its maximum and minimum points would be at +3 and ⁻3.

Tangent is not defined at the values 90 and 270 degrees or $\frac{\pi}{2}$ and $\frac{3\pi}{2}$. Therefore, vertical asymptotes are drawn at those values.

The inverse functions can be graphed in the same manner using a calculator to create a table of values.

TEACHER CERTIFICATION EXAM

COMPETENCY 39.0   ABILITY TO SOLVE PROBLEMS INVOLVING THE SOLUTION OF TRIANGLES.

SKILL 39.1   Solve a triangle, given appropriate parts.

In order to solve a right triangle using trigonometric functions it is helpful to identify the given parts and label them. Usually more than one trigonometric function may be appropriately applied.

Some items to know about right triangles:

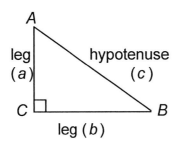

Given angle A, the side labeled leg (a) Is adjacent angle A. And the side labeled leg (b) is opposite to angle A.

Sample problem:

1. Find the missing side.

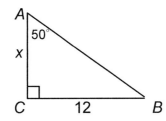

1. Identify the known values. Angle $A = 50$ degrees and the side opposite the given angle is 12. The missing side is the adjacent leg.

$\tan A = \dfrac{\text{opposite}}{\text{adjacent}}$

2. The information suggests the use of the tangent function

3. Write the function.

$\tan 50 = \dfrac{12}{x}$

4. Substitute.

$1.192 = \dfrac{12}{x}$

5. Solve.

$x(1.192) = 12$

$x = 10.069$

Remember that since angle A and angle B are complimentary, then angle $B = 90 - 50$ or 40 degrees.

MATHEMATICS HIGH SCHOOL

Using this information we could have solved for the same side only this time it is the leg opposite from angle B.

$\tan B = \dfrac{\text{opposite}}{\text{adjacent}}$ 1. Write the formula.

$\tan 40 = \dfrac{x}{12}$ 2. Substitute.

$12(.839) = x$ 3. Solve.

$10.069 \approx x$

Now that the two sides of the triangle are known, the third side can be found using the Pythagorean Theorem.

SKILL 39.2 Apply the law of cosines.

Definition: For any triangle ABC, when given two sides and the included angle, the other side can be found using one of the formulas below:

$$a^2 = b^2 + c^2 - (2bc)\cos A$$
$$b^2 = a^2 + c^2 - (2ac)\cos B$$
$$c^2 = a^2 + b^2 - (2ab)\cos C$$

Similarly, when given three sides of a triangle, the included angles can be found using the derivation:

$$\cos A = \dfrac{b^2 + c^2 - a^2}{2bc}$$
$$\cos B = \dfrac{a^2 + c^2 - b^2}{2ac}$$
$$\cos C = \dfrac{a^2 + b^2 - c^2}{2ab}$$

Sample problem:

1. Solve triangle ABC, if angle $B = 87.5°$, $a = 12.3$, and $c = 23.2$. (Compute to the nearest tenth).

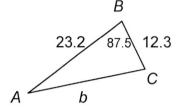

1. Draw and label a sketch.

Find side b.

$b^2 = a^2 + c^2 - (2ac)\cos B$ 2. Write the formula.
$b^2 = (12.3)^2 + (23.2)^2 - 2(12.3)(23.2)(\cos 87.5)$ 3. Substitute.
$b^2 = 664.636$
$b = 25.8$ (rounded)   4. Solve.

Use the law of sine to find angle A.

$\dfrac{\sin A}{a} = \dfrac{\sin B}{b}$ 1. Write formula.

$\dfrac{\sin A}{12.3} = \dfrac{\sin 87.5}{25.8} = \dfrac{.999}{25.8}$ 2. Substitute.

$\sin A = 0.47629$
Angle $A = 28.4$ 3. Solve.

Therefore, angle $C = 180 - (87.5 + 28.4)$
$= 64.1$

2. Solve triangle ABC if $a = 15$, $b = 21$, and $c = 18$. (Round to the nearest tenth).

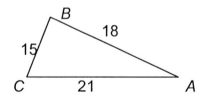

1. Draw and label sketch.

Find angle A.

$\cos A = \dfrac{b^2 + c^2 - a^2}{2bc}$ 2. Write formula.

$\cos A = \dfrac{21^2 + 18^2 - 15^2}{2(21)(18)}$ 3. Substitute.

$\cos A = 0.714$ 4. Solve.
Angle $A = 44.4$

Find angle B.

$\cos B = \dfrac{a^2 + c^2 - b^2}{2ac}$ 5. Write formula.

$\cos B = \dfrac{15^2 + 18^2 - 21^2}{2(15)(18)}$ 6. Substitute.

$\cos B = 0.2$ 7. Solve.
Angle $B = 78.5$

Therefore, angle $C = 180 - (44.4 + 78.5)$
$= 57.1$

SKILL 39.3 Apply the law of sines.

Definition: For any triangle ABC, where a, b, and c are the lengths of the sides opposite angles A, B, and C respectively.

$$\dfrac{\sin A}{a} = \dfrac{\sin B}{b} = \dfrac{\sin C}{c}$$

Sample problem:

1. An inlet is 140 feet wide. The lines of sight from each bank to an approaching ship are 79 degrees and 58 degrees. What are the distances from each bank to the ship?

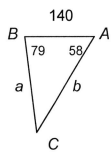

1. Draw and label a sketch.

2. The missing angle is $180 - (79 + 58) = 43$ degrees.

$$\frac{\sin A}{a} = \frac{\sin B}{b} = \frac{\sin C}{c}$$

3. Write formula.

Side opposite 19 degree angle:

$$\frac{\sin 79}{b} = \frac{\sin 43}{140}$$

$$b = \frac{140(.7826)}{.6820}$$

$$b \approx 201.501 \text{ feet}$$

4. Substitute.

5. Solve.

Side opposite 58 degree angle:

$$\frac{\sin 58}{a} = \frac{\sin 43}{140}$$

$$a = \frac{140(.8480)}{.6820}$$

$$a \approx 174.076 \text{ feet}$$

6. Substitute.

7. Solve.

SKILL 39.4 Determine probabilities for a triangle where two sides and a Nonincluded angle are given (ambiguous case).

When the measure of two sides and an angle not included between the sides are given, there are several possible solutions. There may be one, two or no triangles that may be formed from the given information.

The ambiguous case is described using two situations: either angle A is acute or it is obtuse.

Case 1: Angle A is acute.

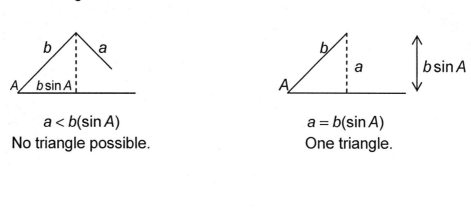

$a < b(\sin A)$
No triangle possible.

$a = b(\sin A)$
One triangle.

$a > b(\sin A)$ and $a \geq b$
One triangle.

$a > b(\sin A)$ but $a < b$
Two triangles.

Case 2: Angle A is obtuse.

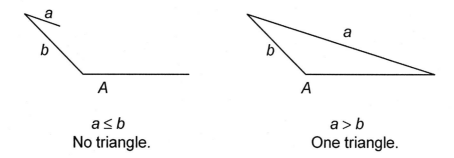

$a \leq b$
No triangle.

$a > b$
One triangle.

Sample problems:

Determine how many solutions exist.

1. Angle $A = 70$ degrees, $a = 16$, $b = 21$.
   Angle A is an acute angle and $a < b$ so we are looking at a case 1 triangle. In this case, $b(\sin A) = 19.734$. Since $a = 16$, this gives us a triangle possibility similar to case 1 triangle 1 where $a < b(\sin A)$ and there are no possible solutions.

2. Angle $C = 95.1$ degrees, $b = 16.8$, and $c = 10.9$.
   Angle C is an obtuse angle and $c \leq b$ (case 2, triangle 1). There are no possible solutions.

3. Angle $B = 45$ degrees, $b = 40$, and $c = 32$.
   Angle B is acute and $b \geq c$. Finding $c(\sin B)$ gives 22.627 and therefore, $b > c(\sin B)$. This indicates a case 1 triangle with one possible solution.

Find the number of possible solutions and then the missing sides, if possible (round all answers to the nearest whole number).

4. Angle $A = 37$ degrees, $a = 49$ and $b = 54$.
   Angle A is acute and $a < b$, find $b(\sin A)$. If $a > b(\sin A)$ there will be two triangles, and if $a < b(\sin A)$ there will be no triangles possible.

   $a = 49$ and $b(\sin A) = 32.498$, therefore, $a > b(\sin A)$ and there are two triangle solutions.

   Use law of sines to find angle B.

   $\dfrac{\sin A}{a} = \dfrac{\sin B}{b}$ 1. Write formula.

   $\dfrac{\sin 37}{49} = \dfrac{\sin B}{54}$ 2. Substitute.

   $\sin B = 0.663$ 3. Solve.

   Angle $B = 42$ degrees or angle $B = 180 - 42 = 138$ degrees. There are two possible solutions to be solved for.

TEACHER CERTIFICATION EXAM

Case 1 (angle $B = 42$ degrees)

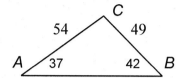

Angle $C = 180 - (37 + 42)$
$\qquad = 101$ degrees

$$\frac{\sin A}{a} = \frac{\sin C}{c}$$

$$\frac{\sin 37}{49} = \frac{\sin 101}{c}$$
$c = 79.924 \approx 80$

1. Draw sketch

2. Find angle C.

3. Find side c using law of sines

4. Substitute.

Case 2 (angle $B = 138$)

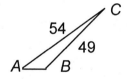

Angle $C = 180 - (37 + 138)$
$\qquad = 5$ degrees

$$\frac{\sin A}{a} = \frac{\sin C}{c}$$

$$\frac{\sin 37}{49} = \frac{\sin 5}{c}$$
$c = 7.097 \approx 7$

1. Draw a sketch.
   ($A = 37$ and $B = 138$)

2. Find angle C.

3. Find side c using law of sine.

4. Substitute.

MATHEMATICS HIGH SCHOOL

When finding the missing sides and angles of triangles that are either acute or obtuse using the law of sines and the law of cosines is imperative. Below is a chart to assist in determining the correct usage.

Given	Suggested Solution Method
(SAS Two sides and the included angle	Law of Cosines will give you the third side. Then use the Law of Sines for the angles.
(SSS) Three sides	Law of Cosines will give you an angle. Then the Law of Sines can be used for the other angles.
(SAA or ASA) One side, two angles	Find the remaining angle and then use the Law of Sines.
(SSA) Two sides, angle not included	Find the number of possible solutions. Use the Law of Sines.

SKILL 39.5 Solve for the area of an oblique triangle, given three appropriate parts.

Definition: The area of any triangle ABC can be found using one of these formulas when given two legs and the included angle:

$$\text{Area} = \frac{1}{2} bc \sin A$$

$$\text{Area} = \frac{1}{2} ac \sin B$$

$$\text{Area} = \frac{1}{2} ab \sin C$$

Sample problem:

Find the area of triangle ABC with $a = 4.2$, $b = 2.6$ and angle $C = 43$ degrees.

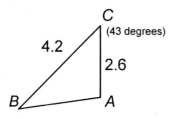

Area = $\frac{1}{2}ab\sin C$ 2. Write the formula.

= $\frac{1}{2}(4.2)(2.6)\sin 43$ 3. Substitute.

= 3.724 square units   4. Solve.

1. Draw and label the sketch.

When only the lengths of the sides are known, it is possible to find the area of the triangle *ABC* using Heron's Formula:

Area = $\sqrt{s(s-a)(s-b)(s-c)}$ where $s = \frac{1}{2}(a+b+c)$

Sample problem:

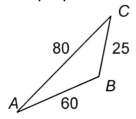

1. Draw and label the sketch.

First, find s.

$s = \frac{1}{2}(a+b+c)$ 2. Write the formula.

$s = \frac{1}{2}(25 + 80 + 60)$ 3. Substitute.

$s = 82.5$ 4. Solve.

Now find the area.

$A = \sqrt{s(s-a)(s-b)(s-c)}$ 5. Write the formula.

$A = \sqrt{82.5(82.5-25)(82.5-80)(82.5-60)}$ 6. Substitute.

$A = 516.5616$ square units   7. Solve.

TEACHER CERTIFICATION EXAM

COMPETENCY 40.0   **UNDERSTANDING THE THEORY OF FUNCTIONS.**

SKILL 40.1   Interpret the language and notation of functions.

A function can be defined as a set of ordered pairs in which each element of the domain is paired with one and only one element of the range. The symbol $f(x)$ is read "f of x." Letter other than "f" can be used to represent a function. The letter "g" is commonly used as in $g(x)$.

Sample problems:

1. Given $f(x) = 4x^2 - 2x + 3$, find $f(^-3)$.

(This question is asking for the range value that corresponds to the domain value of $^-3$).

$$f(x) = 4x^2 - 2x + 3$$
$$f(^-3) = 4(^-3)^2 - 2(^-3) + 3$$
$$f(^-3) = 45$$

1. Replace x with $^-3$.
2. Solve.

2. Find f(3) and f(10), given $f(x) = 7$.

$$f(x) = 7$$
$$(3) = 7$$

1. There are no x values to substitute for. This is your answer.

$$f(x) = 7$$
$$f10) = 7$$

2. Same as above.

Notice that both answers are equal to the constant given.

TEACHER CERTIFICATION EXAM

SKILL 40.2 **Operate with the algebra of functions (operations [+,−,×,÷]), determine composites, and determine inverses.**

If $f(x)$ is a function and the value of 3 is in the domain, the corresponding element in the range would be f(3). It is found by evaluating the function for $x = 3$. The same holds true for adding, subtracting, and multiplying in function form.

The symbol f^{-1} is read "the inverse of f". The $^{-1}$ is not an exponent. The inverse of a function can be found by reversing the order of coordinates in each ordered pair that satisfies the function. Finding the inverse functions means switching the place of x and y and then solving for y.

Sample problem:

1. Find $p(a+1) + 3\{p(4a)\}$ if $p(x) = 2x^2 + x + 1$.

Find $p(a+1)$.

$p(a+1) = 2(a+1)^2 + (a+1) + 1$ Substitute $(a+1)$ for x.
$p(a+1) = 2a^2 + 5a + 4$ Solve.

Find $3\{p(4a)\}$.

$3\{p(4a)\} = 3[2(4a)^2 + (4a) + 1]$ Substitute $(4a)$ for x, multiply by 3.
$3\{p(4a)\} = 96a^2 + 12a + 3$ Solve.

$p(a+1) + 3\{p(4a)\} = 2a^2 + 5a + 4 + 96a^2 + 12a + 3$ Combine like terms.

$p(a+1) + 3\{p(4a)\} = 98a^2 + 17a + 7$

MATHEMATICS HIGH SCHOOL

SKILL 40.3 Identify the properties of functions (symmetry and even/odd).

Definition: A function f is even if $f(^-x) = f(x)$ and odd if $f(^-x) = {^-f(x)}$ for all x in the domain of f.

Sample problems:

Determine if the given function is even, odd, or neither even nor odd.

1. $f(x) = x^4 - 2x^2 + 7$
   $f(^-x) = (^-x)^4 - 2(^-x)^2 + 7$
   $f(^-x) = x^4 - 2x^2 + 7$

   $f(x)$ is an even function.

   1. Find $f(^-x)$.
   2. Replace x with ^-x.
   3. Since $f(^-x) = f(x)$, $f(x)$ is an even function.

2. $f(x) = 3x^3 + 2x$
   $f(^-x) = 3(^-x)^3 + 2(^-x)$
   $f(^-x) = {^-3x^3} - 2x$

   $^-f(x) = {^-(3x^3 + 2x)}$
   $^-f(x) = {^-3x^3} - 2x$

   $f(x)$ is an odd function.

   1. Find $f(^-x)$.
   2. Replace x with ^-x.
   3. Since $f(x)$ is not equal to $f(^-x)$, $f(x)$ is not an even function.
   4. Try $^-f(x)$.
   5. Since $f(^-x) = {^-f(x)}$, $f(x)$ is an odd function.

3. $g(x) = 2x^2 - x + 4$
   $g(^-x) = 2(^-x)^2 - (^-x) + 4$
   $g(^-x) = 2x^2 + x + 4$

   $^-g(x) = {^-(2x^2 - x + 4)}$
   $^-g(x) = {^-2x^2} + x - 4$

   $g(x)$ is neither even nor odd.

   1. First find $g(^-x)$.
   2. Replace x with ^-x.
   3. Since $g(x)$ does not equal $g(^-x)$, $g(x)$ is not an even function.
   4. Try $^-g(x)$.
   5. Since $^-g(x)$ does not equal $g(^-x)$, $g(x)$ is not an odd function.

TEACHER CERTIFICATION EXAM

COMPETENCY 41.0   **KNOWLEDGE OF CIRCULAR/TRIGONOMETRIC FUNCTIONS.**

SKILL 41.1 Graph in the polar plane.

One way to graph points is in the rectangular coordinate system. In this system, the point (a,b) describes the point whose distance along the x-axis is "a" and whose distance along the y-axis is "b." The other method used to locate points is the **polar plane coordinate system**. This system consists of a fixed point called the pole or origin (labeled O) and a ray with O as the initial point called the polar axis. The ordered pair of a point P in the polar coordinate system is (r,θ), where $|r|$ is the distance from the pole and θ is the angle measure from the polar axis to the ray formed by the pole and point P. The coordinates of the pole are $(0,\theta)$, where θ is arbitrary. Angle θ can be measured in either degrees or in radians.

Sample problem:

1. Graph the point P with polar coordinates $(^-2, ^-45 \text{ degrees})$.

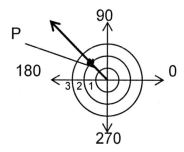

Draw $\theta = ^-45$ degrees in standard position. Since r is negative, locate the point $|^-2|$ units from the pole on the ray opposite the terminal side of the angle. Note that P can be represented by
$(^-2, ^-45 \text{ degrees} + 180 \text{ degrees}) = (2, 135 \text{ degrees})$ or by
$(^-2, ^-45 \text{ degrees} - 180 \text{ degrees}) = (2, ^-225 \text{ degrees})$.

2. Graph the point $P = \left(3, \dfrac{\pi}{4}\right)$ and show another graph that also represents the same point P.

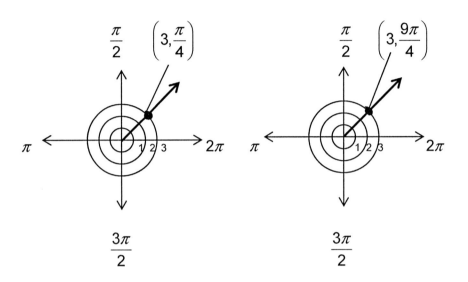

In the second graph, the angle 2π is added to $\dfrac{\pi}{4}$ to give the point $\left(3, \dfrac{9\pi}{4}\right)$.

It is possible that r be allowed to be negative. Now instead of measuring $|r|$ units along the terminal side of the angle, we would locate the point $|{-3}|$ units from the pole on the ray opposite the terminal side. This would give the points $\left(-3, \dfrac{5\pi}{4}\right)$ and $\left(-3, \dfrac{-3\pi}{4}\right)$.

TEACHER CERTIFICATION EXAM

COMPETENCY 42.1   MASTERY IN PERFORMING OPERATIONS ON VECTORS.

SKILL 42.1   Resolve a vector into component vectors.

Occasionally, it is important to reverse the addition or subtraction process and express the single vector as the sum or difference of two other vectors. It may be critically important for a pilot to understand not only the air velocity but also the ground speed and the climbing speed.

Sample problem:

A pilot is traveling at an air speed of 300 mph and a direction of 20 degrees. Find the horizontal vector (ground speed) and the vertical vector (climbing speed).

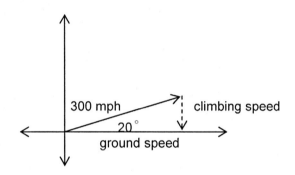

1. Draw sketch.
2. Use appropriate trigonometric ratio to calculate the component vectors.

To find the vertical vector:

$$\sin x = \frac{\text{opposite}}{\text{hypotenuse}}$$

$$\sin(20) = \frac{c}{300}$$
$$c = (.3420)(300)$$
$$c = 102.606$$

To find the horizontal vector:

$$\cos x = \frac{\text{adjacent}}{\text{hypotenuse}}$$

$$\cos(20) = \frac{g}{300}$$
$$g = (.9397)(300)$$
$$g = 281.908$$

SKILL 42.2 Add or subtract vectors.

Vectors are used to measure displacement of an object or force.

Addition of vectors:

$$(a,b)+(c,d)=(a+c,b+d)$$

Addition Properties of vectors:

$$a+b=b+a$$
$$a+(b+c)=(a+b)+c$$
$$a+0=a$$
$$a+(^-a)=0$$

Subtraction of vectors:

$a-b=a+(^-b)$ therefore,
$$a-b=(a_1,a_2)+(^-b_1,^-b_2) \text{ or}$$
$$a-b=(a_1-b_1,a_2-b_2)$$

Sample problem:

If $a=(4,^-1)$ and $b=(^-3,6)$, find $a+b$ and $a-b$.

Using the rule for addition of vectors:

$$(4,^-1)+(^-3,6)=(4+(^-3),^-1+6)$$
$$=(1,5)$$

Using the rule for subtraction of vectors:

$$(4,^-1)-(^-3,6)=(4-(^-3),^-1-6)$$
$$=(7,^-7)$$

TEACHER CERTIFICATION EXAM

SKILL 42.3 Find the dot product of two vectors.

The dot product $a \cdot b$:

$$a = (a_1, a_2) = a_1 i + a_2 j \quad \text{and} \quad b = (b_1, b_2) = b_1 i + b_2 j$$

$$a \cdot b = a_1 b_1 + a_2 b_2$$

$a \cdot b$ is read "a dot b". Dot products are also called scalar or inner products. When discussing dot products, it is important to remember that "a dot b" is not a vector, but a real number.

Properties of the dot product:

$a \cdot a = |a|^2$
$a \cdot b = b \cdot a$
$a \cdot (b + c) = a \cdot b + a \cdot c$
$(ca) \cdot b = c(a \cdot b) = a \cdot (cb)$
$0 \cdot a = 0$

Sample problems:

Find the dot product.

1. $a = (5, 2), b = (^-3, 6)$
   $a \cdot b = (5)(^-3) + (2)(6)$
   $= {}^-15 + 12$
   $= {}^-3$

2. $a = (5i + 3j), b = (4i - 5j)$
   $a \cdot b = (5)(4) + (3)(^-5)$
   $= 20 - 15$
   $= 5$

3. The magnitude and direction of a constant force are given by $a = 4i + 5j$. Find the amount of work done if the point of application of the force moves from the origin to the point $P(7, 2)$.

The work W done by a constant force a as its point of application moves along a vector b is $W = a \cdot b$.

Sketch the constant force vector a and the vector b.

MATHEMATICS HIGH SCHOOL

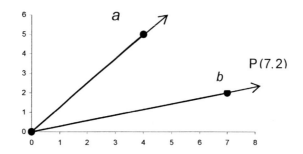

$b = (7,2) = 7i + 2j$

Use the definition of work done to solve.

$$W = a \cdot b$$
$$= (4i + 5j)(7i + 2j)$$
$$= (4)(7) + (5)(2)$$
$$= (28) + (10)$$
$$= 38$$

TEACHER CERTIFICATION EXAM

SKILL 42.4 Solve problems involving vectors.

Vectors are used often in navigation of ships and aircraft as well as in force and work problems.

Sample problem:

1. An airplane is flying with a heading of 60 degrees east of north at 450 mph. A wind is blowing at 37 mph from the north. Find the plane's ground speed to the nearest mph and direction to the nearest degree.

Draw a sketch.

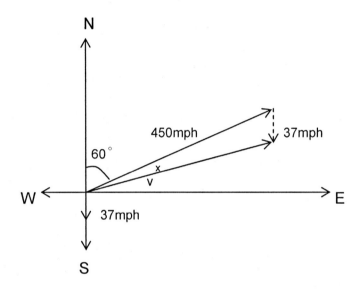

Use the law of cosines to find the ground speed.

$$a^2 = b^2 + c^2 - 2bc \cos A$$
$$|v|^2 = 450^2 + 37^2 - 2(450)(37)\cos 60$$
$$|v|^2 = 187,219$$
$$|v| = 432.688$$
$$|v| \approx 433 \text{ mph}$$

MATHEMATICS HIGH SCHOOL

Use the law of sines to find the measure of x.

$$\frac{\sin A}{a} = \frac{\sin C}{c}$$

$$\frac{\sin 60}{433} = \frac{\sin x}{37}$$

$$\sin x = \frac{37(\sin 60)}{433}$$

$$x = 4.24$$

$$x \approx 4 \text{ degrees}$$

The plane's actual course is $60 + 4 = 64$ degrees east of north and the ground speed is approximately 433 miles per hour.

TEACHER CERTIFICATION EXAM

COMPETENCY 43.0   **PROFICIENCY IN DETERMINING DISTANCES ON A PLANE.**

SKILL 43.1   Determine the distance between a point and a line.

In order to accomplish the task of finding the distance from a given point to another given line the perpendicular line that intersects the point and line must be drawn and the equation of the other line written. From this information the point of intersection can be found. This point and the original point are used in the distance formula given below:

$$D = \sqrt{(x_2 - x_1)^2 + (y_2 - y_1)^2}$$

Sample Problem:

1. Given the point $(^-4, 3)$ and the line $y = 4x + 2$, find the distance from the point to the line.

$y = 4x + 2$	1. Find the slope of the given line by solving for y.
$y = 4x + 2$	2. The slope is $4/1$, the perpendicular line will have a slope of $^-1/4$.
$y = (^-1/4)x + b$	3. Use the new slope and the given point to find the equation of the perpendicular line.
$3 = (^-1/4)(^-4) + b$	4. Substitute $(^-4, 3)$ into the equation.
$3 = 1 + b$	5. Solve.
$2 = b$	6. Given the value for b, write the equation of the perpendicular line.
$y = (^-1/4)x + 2$	7. Write in standard form.
$x + 4y = 8$	8. Use both equations to solve by elimination to get the point of intersection.
$^-4x + y = 2$	
$x + 4y = 8$	9. Multiply the bottom row by 4.

MATHEMATICS HIGH SCHOOL

$$-4x + y = 2$$
$$4x + 16y = 32$$
$$\overline{}$$
$$17y = 34$$
$$y = 2$$

10. Solve.

$$y = 4x + 2$$
$$2 = 4x + 2$$
$$x = 0$$

11. Substitute to find the x value.
12. Solve.

(0,2) is the point of intersection. Use this point on the original line and the original point to calculate the distance between them.

$$D = \sqrt{(x_2 - x_1)^2 + (y_2 - y_1)^2} \quad \text{where points are (0,2) and (-4,3).}$$

$$D = \sqrt{(-4 - 0)^2 + (3 - 2)^2} \qquad \text{1. Substitute.}$$

$$D = \sqrt{(16) + (1)} \qquad \text{2. Simplify.}$$

$$D = \sqrt{17}$$

TEACHER CERTIFICATION EXAM

SKILL 43.2 Determine the distance between two parallel lines.

The distance between two parallel lines, such as line *AB* and line *CD* as shown below is the line segment *RS*, the perpendicular between the two parallels.

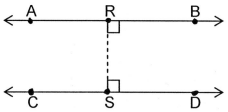

Sample Problem:

Given the geometric figure below, find the distance between the two parallel sides *AB* and *CD*.

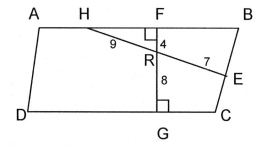

The distance *FG* is 12 units.

SKILL 43.3 Apply the distance formula.

The key to applying the distance formula is to understand the problem before beginning.

$$D = \sqrt{(x_2 - x_1)^2 + (y_2 - y_1)^2}$$

Sample Problem:

1. Find the perimeter of a figure with vertices at $(4,5)$, $(^-4,6)$ and $(^-5,^-8)$.

The figure being described is a triangle. Therefore, the distance for all three sides must be found. Carefully, identify all three sides before beginning.

Side 1 = $(4,5)$ to $(^-4,6)$
Side 2 = $(^-4,6)$ to $(^-5,^-8)$
Side 3 = $(^-5,^-8)$ to $(4,5)$

$$D_1 = \sqrt{(^-4-4)^2 + (6-5)^2} = \sqrt{65}$$

$$D_2 = \sqrt{(^-5-(^-4))^2 + (^-8-6)^2} = \sqrt{197}$$

$$D_3 = \sqrt{(4-(^-5))^2 + (5-(^-8))^2} = \sqrt{250} \text{ or } 5\sqrt{10}$$

$$\text{Perimeter} = \sqrt{65} + \sqrt{197} + 5\sqrt{10}$$

TEACHER CERTIFICATION EXAM

SKILL 43.4 Apply the formula for midpoint.

Midpoint Definition:

If a line segment has endpoints of (x_1, y_1) and (x_2, y_2), then the midpoint can be found using:

$$\left(\frac{x_1 + x_2}{2}, \frac{y_1 + y_2}{2} \right)$$

Sample problems:

1. Find the center of a circle with a diameter whose endpoints are (3,7) and (−4, −5).

$$\text{Midpoint} = \left(\frac{3 + (^-4)}{2}, \frac{7 + (^-5)}{2} \right)$$

$$\text{Midpoint} = \left(\frac{^-1}{2}, 1 \right)$$

2. Find the midpoint given the two points $\left(5, 8\sqrt{6}\right)$ and $\left(9, ^-4\sqrt{6}\right)$.

$$\text{Midpoint} = \left(\frac{5+9}{2}, \frac{8\sqrt{6} + (^-4\sqrt{6})}{2} \right)$$

$$\text{Midpoint} = \left(7, 2\sqrt{6}\right)$$

COMPETENCY 44.0   **COMPREHENSION OF SYSTEMS OF HIGHER DEGREE EQUATIONS.**

SKILL 44.1   **Solve systems of higher degree equations.**

Systems of quadratic equations can be solved by graphing but this method is sometimes inconclusive depending on the accuracy of the graphs and can also be cumbersome. It is recommended that either a substitution or elimination method be considered.

Sample problems:

Find the solution to the system of equations.

1. $y^2 - x^2 = {}^-9$
   $2y = x - 3$

   1. Use substitution method solving the second equation for x.

   $2y = x - 3$
   $x = 2y + 3$

   2. Substitute this into the first equation in place of (x).

   $y^2 - (2y + 3)^2 = {}^-9$
   $y^2 - (4y^2 + 12y + 9) = {}^-9$
   $y^2 - 4y^2 - 12y - 9 = {}^-9$
   ${}^-3y^2 - 12y - 9 = {}^-9$
   ${}^-3y^2 - 12y = 0$
   ${}^-3y(y + 4) = 0$

   3. Solve.

   4. Factor.

   ${}^-3y = 0$ $y + 4 = 0$
   $y = 0$ $y = {}^-4$

   5. Set each factor equal to zero.
   6. Use these values for y to solve for x.

   $2y = x - 3$ $2y = x - 3$
   $2(0) = x - 3$ $2({}^-4) = x - 3$
   $0 = x - 3$ ${}^-8 = x - 3$
   $x = 3$ $x = {}^-5$

   7. Choose an equation.
   8. Substitute.
   9. Write ordered pairs.

   $(3, 0)$ and $({}^-5, {}^-4)$ satisfy the system of equations given.

2. $\begin{array}{l}{}^-9x^2+y^2=16\\5x^2+y^2=30\end{array}$ 　　　　Use elimination to solve.

$$\begin{array}{l}{}^-9x^2+y^2=16\\\underline{{}^-5x^2-y^2={}^-30}\\{}^-14x^2={}^-14\\x^2=1\\x=\pm 1\end{array}$$

1. Multiply second row by $^-1$.
2. Add.
3. Divide by $^-14$.
4. Take the square root of both sides

$\begin{array}{ll}{}^-9(1)^2+y^2=16 & {}^-9({}^-1)^2+y^2=16\\{}^-9+y^2=16 & {}^-9+y^2=16\\y^2=25 & y^2=25\\y=\pm 5 & y=\pm 5\\(1,\pm 5) & ({}^-1,\pm 5)\end{array}$

5. Substitute both values of x into the equation.
6. Take the square root of both sides.
7. Write the ordered pairs.

TEACHER CERTIFICATION EXAM

COMPETENCY 45.0   ABILITY TO DETERMINE THE EQUATIONS OF LOCI.

SKILL 45.1 Determine the equation locus, given its description.

A locus is the set of all points that satisfy a given condition. The solution to a locus problem is both a figure showing the locus and a statement or equation that names the locus and defines its position.

Sample Problem:

What is the locus of points 6 cm from a fixed point (0,0) in a plane?

1. Draw figure.

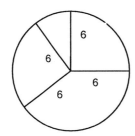

2. Write statement to name locus and define its position.

The locus is a circle with (0,0) as the center and a radius of 6 centimeters. The equation of this circle would be $x^2 + y^2 = 36$.

Practice problem:

1. Describe the locus of points equidistant from two given parallel lines $y = 3x + 1$ and $^-3x + y = {}^-5$.

SKILL 45.2 Identify the directix, foci, vertices, and axes of a conic section where appropriate.

PARABOLAS - A parabola is a set of all points in a plane that are equidistant from a fixed point (focus) and a line (directrix).

FORM OF EQUATION	$y = a(x-h)^2 + k$	$x = a(y-k)^2 + h$
IDENTIFICATION	x^2 term, y not squared	y^2 term, x not squared

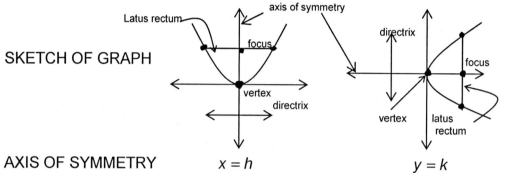

SKETCH OF GRAPH

AXIS OF SYMMETRY	$x = h$	$y = k$

-A line through the vertex and focus upon which the parabola is symmetric.

VERTEX	(h, k)	(h, k)

-The point where the parabola intersects the axis of symmetry.

FOCUS	$(h, k + 1/4a)$	$(h + 1/4a, k)$
DIRECTRIX	$y = k - 1/4a$	$x = h - 1/4a$
DIRECTION OF OPENING	up if $a > 0$, down if $a < 0$	right if $a > 0$, left if $a < 0$
LENGTH OF LATUS RECTUM	$\lvert 1/a \rvert$	$\lvert 1/a \rvert$

-A chord through the focus, perpendicular to the axis of symmetry, with endpoints on the parabola.

Sample Problem:

1. Find all identifying features of $y = {}^-3x^2 + 6x - 1$.

First, the equation must be put into the general form $y = a(x-h)^2 + k$.

$y = {}^-3x^2 + 6x - 1$ 1. Begin by completing the square.
$ = {}^-3(x^2 - 2x + 1) - 1 + 3$
$ = {}^-3(x-1)^2 + 2$ 2. Using the general form of the equation begin to identify known variables.

$a = {}^-3 \quad h = 1 \quad k = 2$

axis of symmetry: $x = 1$
vertex: $(1, 2)$
focus: $(1, 1\ 11/12)$
directrix: $y = 2\ 1/12$
direction of opening: down since $a < 0$
length of latus rectum: $1/3$

ELLIPSE

FORM OF EQUATION	$\dfrac{(x-h)^2}{a^2}+\dfrac{(y-k)^2}{b^2}=1$	$\dfrac{(x-h)^2}{b^2}+\dfrac{(y-k)^2}{a^2}=1$
(for ellipses where $a^2 > b^2$).	where $b^2 = a^2 - c^2$	where $b^2 = a^2 - c^2$
IDENTIFICATION	horizontal major axis	vertical major axis

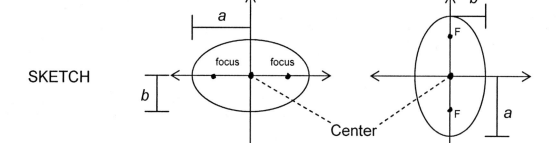

SKETCH

CENTER	(h,k)	(h,k)
FOCI	$(h \pm c, k)$	$(h, k \pm c)$
MAJOR AXIS LENGTH	$2a$	$2a$
MINOR AXIS LENGTH	$2b$	$2b$

Sample Problem:

Find all identifying features of the ellipse $2x^2 + y^2 - 4x + 8y - 6 = 0$.

First, begin by writing the equation in standard form for an ellipse.

$2x^2 + y^2 - 4x + 8y - 6 = 0$ 1. Complete the square for each variable.

$2(x^2 - 2x + 1) + (y^2 + 8y + 16) = 6 + 2(1) + 16$

$2(x-1)^2 + (y+4)^2 = 24$ 2. Divide both sides by 24.

$\dfrac{(x-1)^2}{12} + \dfrac{(y+4)^2}{24} = 1$ 3. Now the equation is in standard form.

MATHEMATICS HIGH SCHOOL 247

Identify known variables: $h = 1 \quad k = {}^-4 \quad a = \sqrt{24} \text{ or } 2\sqrt{6}$
$b = \sqrt{12} \text{ or } 2\sqrt{3} \quad c = 2\sqrt{3}$

Identification: vertical major axis
Center: $(1, {}^-4)$
Foci: $(1, {}^-4 \pm 2\sqrt{3})$
Major axis: $4\sqrt{6}$
Minor axis: $4\sqrt{3}$

HYPERBOLA

FORM OF EQUATION	$\dfrac{(x-h)^2}{a^2} - \dfrac{(y-k)^2}{b^2} = 1$ where $c^2 = a^2 + b^2$	$\dfrac{(y-k)^2}{a^2} - \dfrac{(x-h)^2}{b^2} = 1$ where $c^2 = a^2 + b^2$
IDENTIFICATION	horizontal transverse axis (y^2 is negative)	vertical transverse axis (x^2 is negative)
SKETCH	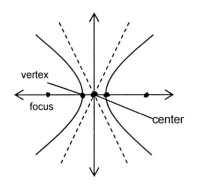	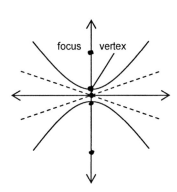
SLOPE OF ASYMPTOTES	$\pm(b/a)$	$\pm(b/a)$
TRANSVERSE AXIS (endpoints are vertices of the hyperbola and goes through center)	$2a$	$2a$
CONJUGATE AXIS (perpendicular to transverse axis at center)	$2b$	$2b$
CENTER	(h,k)	(h,k)
FOCI	$(h \pm c, k)$	$(h, k \pm c)$
VERTICES	$(h \pm a, k)$	$(h, k \pm a)$

Sample Problem:

Find all the identifying features of a hyperbola given its equation.

$$\frac{(x+3)^2}{4} - \frac{(y-4)^2}{16} = 1$$

Identify all known variables: $h = {}^-3 \quad k = 4 \quad a = 2 \quad b = 4 \quad c = 2\sqrt{5}$

Slope of asymptotes: $\pm 4/2$ or ± 2
Transverse axis: 4 units long
Conjugate axis: 8 units long
Center: $({}^-3, 4)$
Foci: $({}^-3 \pm 2\sqrt{5}, 4)$
Vertices: $({}^-1, 4)$ and $({}^-5, 4)$

COMPETENCY 46.0 KNOWLEDGE OF EQUATIONS OF CONIC SECTIONS AND THEIR USE.

SKILL 46.1 Determine the center of and the radius and graph a circle, given its equation.

The equation of a circle with its center at (h,k) and a radius r units is:

$$(x-h)^2 + (y-k)^2 = r^2$$

Sample Problem:

1. Given the equation $x^2 + y^2 = 9$, find the center and the radius of the circle. Then graph the equation.

First, writing the equation in standard circle form gives:

$$(x-0)^2 + (y-0)^2 = 3^2$$

therefore, the center is (0,0) and the radius is 3 units.

Sketch the circle:

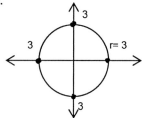

2. Given the equation $x^2 + y^2 - 3x + 8y - 20 = 0$, find the center and the radius. Then graph the circle.

First, write the equation in standard circle form by completing the square for both variables.

$x^2 + y^2 - 3x + 8y - 20 = 0$ 1. Complete the squares.

$(x^2 - 3x + 9/4) + (y^2 + 8y + 16) = 20 + 9/4 + 16$
$(x - 3/2)^2 + (y + 4)^2 = 153/4$

The center is $(3/2, ^-4)$ and the radius is $\sqrt{153}/2$ or $(3\sqrt{17})/2$.

Graph the circle.

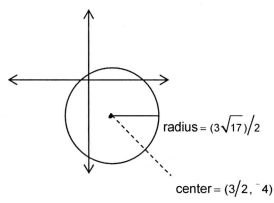

radius = $(3\sqrt{17})/2$

center = $(3/2, {}^-4)$

SKILL 46.2 Identify the equation of a circle, given its center and radius.

To write the equation given the center and the radius use the standard form of the equation of a circle:

$$(x - h)^2 + (y - k)^2 = r^2$$

Sample problems:

Given the center and radius, write the equation of the circle.

1. Center $({}^-1, 4)$; radius 11

   $(x - h)^2 + (y - k)^2 = r^2$ 1. Write standard equation.
   $(x - ({}^-1))^2 + (y - (4))^2 = 11^2$ 2. Substitute.
   $(x + 1)^2 + (y - 4)^2 = 121$ 3. Simplify.

2. Center $(\sqrt{3}, {}^-1/2)$; radius $= 5\sqrt{2}$

   $(x - h)^2 + (y - k)^2 = r^2$ 1. Write standard equation.
   $(x - \sqrt{3})^2 + (y - ({}^-1/2))^2 = (5\sqrt{2})^2$ 2. Substitute.
   $(x - \sqrt{3})^2 + (y + 1/2)^2 = 50$ 3. Simplify.

COMPETENCY 47.0   **PROFICIENCY IN IDENTIFYING AND GRAPHING POLYNOMIAL AND RATIONAL FUNCTIONS.**

SKILL 47.1   Determine and identify the intercepts and asymptotes of a given rational function.

A rational function is given in the form $f(x) = p(x)/q(x)$. In the equation, $p(x)$ and $q(x)$ both represent polynomial functions where $q(x)$ does not equal zero. The branches of rational functions approach asymptotes. Setting the denominator equal to zero and solving will give the value(s) of the vertical asymptotes(s) since the function will be undefined at this point. If the value of $f(x)$ approaches b as the $|x|$ increases, the equation $y = b$ is a horizontal asymptote. To find the horizontal asymptote it is necessary to make a table of value for x that are to the right and left of the vertical asymptotes. The pattern for the horizontal asymptotes will become apparent as the $|x|$ increases.

If there are more than one vertical asymptotes, remember to choose numbers to the right and left of each one in order to find the horizontal asymptotes and have sufficient points to graph the function.

Sample problem:

1. Graph $f(x) = \dfrac{3x+1}{x-2}$.

$x - 2 = 0$
$x = 2$

1. Set denominator $=0$ to find the vertical asymptote.

x	$f(x)$
3	10
10	3.875
100	3.06
1000	3.006
1	$^-4$
$^-10$	2.417
$^-100$	2.93
$^-1000$	2.99

2. Make table choosing numbers to the right and left of the vertical asymptote.

3. Tthe pattern shows that as the $|x|$ increases $f(x)$ approaches the value 3, therefore a horizontal asymptote exists at $y = 3$

Sketch the graph.

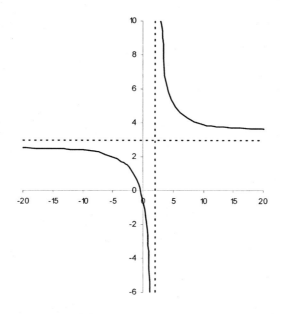

COMPETENCY 48.0 UNDERSTANDING OF THE CONCEPTS OF LOGARITHMIC AND EXPONENTIAL FUNCTIONS.

SKILL 48.1 Convert natural or common logarithmic functions into exponential functions and vice versa.

When changing common logarithms to exponential form,

$$y = \log_b x \quad \text{if and only if} \quad x = b^y$$

Natural logarithms can be changed to exponential form by using,

$$\log_e x = \ln x \quad \text{or} \quad \ln x = y \quad \text{can be written as} \quad e^y = x$$

Practice Problems:

Express in exponential form.

1. $\log_3 81 = 4$
   $x = 81 \quad b = 3 \quad y = 4$ Identify values.
   $81 = 3^4$ Rewrite in exponential form.

Solve by writing in exponential form.

2. $\log_x 125 = 3$

   $x^3 = 125$ Write in exponential form.
   $x^3 = 5^3$ Write 125 in exponential form.
   $x = 5$ Bases must be equal if exponents are equal.

Use a scientific calculator to solve.

3. Find $\ln 72$.
   $\ln 72 = 4.2767$ Use the $\ln x$ key to find natural logs.

4. Find $\ln x = 4.2767$ Write in exponential form.
   $e^{4.2767} = x$ Use the key (or 2nd $\ln x$) to find x.
   $x = 72.002439$ The small difference is due to rounding.

SKILL 48.2 Apply the properties of logarithmic or exponential functions to solve equations.

To solve logarithms or exponential functions it is necessary to use several properties.

Multiplication Property $\quad\log_b mn = \log_b m + \log_b n$

Quotient Property $\quad\log_b \dfrac{m}{n} = \log_b m - \log_b n$

Powers Property $\quad\log_b n^r = r \log_b n$

Equality Property $\quad\log_b n = \log_b m \quad$ if and only if $n = m$.

Change of Base Formula $\quad\log_b n = \dfrac{\log n}{\log b}$

$\log_b b^x = x$ and $b^{\log_b x} = x$

Sample problem.

Solve for x.
1. $\log_6(x-5) + \log_6 x = 2$

   $\log_6 x(x-5) = 2$ Use product property.

   $\log_6 x^2 - 5x = 2$ Distribute.

   $x^2 - 5x = 6^2$ Write in exponential form.

   $x^2 - 5x - 36 = 0$ Solve quadratic equation.

   $(x+4)(x-9) = 0$

   $x = {}^-4 \quad x = 9$

***Be sure to check results. Remember x must be greater than zero in $\log x = y$.

Check: $\log_6(x-5) + \log_6 x = 2$

$\log_6(^-4-5) + \log_6(^-4) = 2$ Substitute the first answer $^-4$.

$\log_6(^-9) + \log_6(^-4) = 2$ This is undefined, x is less than zero.

$\log_6(9-5) + \log_6 9 = 2$ Substitute the second answer 9.

$\log_6 4 + \log_6 9 = 2$

$\log_6(4)(9) = 2$ Multiplication property.

$\log_6 36 = 2$

$6^2 = 36$ Write in exponential form.

$36 = 36$

Practice problems:

1. $\log_4 x = 2\log_4 3$

2. $2\log_3 x = 2 + \log_3(x-2)$

3. Use change of base formula to find $(\log_3 4)(\log_4 3)$.

TEACHER CERTIFICATION EXAM

COMPETENCY 49.0   **ABILITY TO APPLY THE CONCEPT OF LIMITS TO FUNCTIONS.**

SKILL 49.1   **Solve problems using the laws of limits concerning sums, products, and quotients of functions.**

The limit of a function is the y value that the graph approaches as the x values approach a certain number. To find a limit there are two points to remember.

1. Factor the expression completely and cancel all common factors in fractions.
2. Substitute the number to which the variable is approaching. In most cases this produces the value of the limit.

If the variable in the limit is approaching ∞, factor and simplify first; then examine the result. If the result does not involve a fraction with the variable in the denominator, the limit is usually also equal to ∞. If the variable is in the denominator of the fraction, the denominator is getting larger which makes the entire fraction smaller. In other words the limit is zero.

Examples:

1. $\lim\limits_{x \to ^-3} \dfrac{x^2 + 5x + 6}{x + 3} + 4x$ Factor the numerator.

$\lim\limits_{x \to ^-3} \dfrac{(x+3)(x+2)}{(x+3)} + 4x$ Cancel the common factors.

$\lim\limits_{x \to ^-3} (x+2) + 4x$ Substitute $^-3$ for x.

$(^-3 + 2) + 4(^-3)$ Simplify.

$^-1 + ^-12$

$^-13$

2. $\lim\limits_{x \to \infty} \dfrac{2x^2}{x^5}$ Cancel the common factors.

$\lim\limits_{x \to \infty} \dfrac{2}{x^3}$ Since the denominator is getting larger, the entire fraction is getting smaller. The fraction is getting close to zero.

$\dfrac{2}{\infty^3}$

0

MATHEMATICS HIGH SCHOOL   257

Practice problems:

1. $\lim\limits_{x \to \pi} 5x^2 + \sin x$

2. $\lim\limits_{x \to {}^-4} \dfrac{x^2 + 9x + 20}{x + 4}$

SKILL 49.2 Determine the limit of a quotient of functions using L'Hopital's rule.

After simplifying an expression to evaluate a limit, substitute the value that the variable approaches. If the substitution results in either $0/0$ or ∞/∞, use L'Hopital's rule to find the limit.

L'Hopital's rule states that you can find such limits by taking the derivative of the numerator and the derivative of the denominator, and then finding the limit of the resulting quotient.
(For hints on taking derivatives, see section 50.0.)

Examples:

1. $\lim\limits_{x \to \infty} \dfrac{3x - 1}{x^2 + 2x + 3}$ No factoring is possible.

$\dfrac{3\infty - 1}{\infty^2 + 2\infty + 3}$ Substitute ∞ for x.

$\dfrac{\infty}{\infty}$ Since a constant times infinity is still a large number, $3(\infty) = \infty$.

$\lim\limits_{x \to \infty} \dfrac{3}{2x + 2}$ To find the limit, take the derivative of the numerator and denominator.

$\dfrac{3}{2(\infty) + 2}$ Substitute ∞ for x again.

$\dfrac{3}{\infty}$ Since the denominator is a very large

0 number, the fraction is getting smaller. Thus the limit is zero.

2. $\lim_{x \to 1} \dfrac{\ln x}{x-1}$ Substitute 1 for x.

   $\dfrac{\ln 1}{1-1}$ The $\ln 1 = 0$

   $\dfrac{0}{0}$ To find the limit, take the derivative of the numerator and denominator.

   $\lim_{x \to 1} \dfrac{\frac{1}{x}}{1}$ Substitute 1 for x again.

   $\dfrac{\frac{1}{1}}{1}$ Simplify. The limit is one.

   1

Practice problems:

1. $\lim_{x \to \infty} \dfrac{x^2 - 3}{x}$

2. $\lim_{x \to \frac{\pi}{2}} \dfrac{\cos x}{x - \frac{\pi}{2}}$

TEACHER CERTIFICATION EXAM

COMPETENCY 50.0   **ABILITY TO FIND DERIVATIVES OF FUNCTIONS**

SKILL 50.1   **Find the derivatives of algebraic functions.**

    A. Derivative of a constant--for any constant, the derivative is always zero.

    B. Derivative of a variable--the derivative of a variable (i.e. x) is one.

    C. Derivative of a variable raised to a power--for variable expressions with rational exponents (i.e. $3x^2$) multiply the coefficient (3) by the exponent (2) then subtract one (1) from the exponent to arrive at the derivative ($6x^{2-1} = 6x$).

Example:

1. $y = 5x^4$ Take the derivative.

   $\dfrac{dy}{dx} = (5)(4)x^{4-1}$ Multiply the coefficient by the exponent and subtract 1 from the exponent.

   $\dfrac{dy}{dx} = 20x^3$ Simplify.

2. $y = \dfrac{1}{4x^3}$ Rewrite using negative exponent.

   $y = \dfrac{1}{4}x^{-3}$ Take the derivative.

   $\dfrac{dy}{dx} = \left(\dfrac{1}{4} \cdot {-3}\right)x^{-3-1}$

   $\dfrac{dy}{dx} = \dfrac{-3}{4}x^{-4} = \dfrac{-3}{4x^4}$ Simplify.

3. $y = 3\sqrt{x^5}$ Rewrite using $\sqrt[z]{x^n} = x^{n/z}$.

   $y = 3x^{5/2}$ Take the derivative.

   $\dfrac{dy}{dx} = (3)\left(\dfrac{5}{2}\right)x^{5/2-1}$

   $\dfrac{dy}{dx} = \left(\dfrac{15}{2}\right)x^{3/2}$ Simplify.

   $\dfrac{dy}{dx} = 17.5\sqrt{x^3} = 17.5x\sqrt{x}$

MATHEMATICS HIGH SCHOOL

TEACHER CERTIFICATION EXAM

SKILL 50.2 Find the derivatives of trigonometric functions.

    A. sin x --the derivative of the sine of x is simply the cosine of x.

    B. cos x --the derivative of the cosine of x is negative one ($^-1$) times the sine of x.

    C. tan x --the derivative of the tangent of x is the square of the secant of x.

If object of the trig. function is an expression other than x, follow the above rules substituting the expression for x. The only additional step is to multiply the result by the derivative of the expression.

Examples:

1. $y = \pi \sin x$ Carry the coefficient (π) throughout the problem.

         $\dfrac{dy}{dx} = \pi \cos x$

2. $y = \dfrac{2}{3} \cos x$ Do not forget to multiply the coefficient by negative one when taking the

         $\dfrac{dy}{dx} = \dfrac{^-2}{3} \sin x$ derivative of a cosine function.

3. $y = 4 \tan(5x^3)$

         $\dfrac{dy}{dx} = 4 \sec^2(5x^3)(5 \bullet 3x^{3-1})$ The derivative of tan x is sec$^2 x$.

         $\dfrac{dy}{dx} = 4 \sec^2(5x^3)(15x^2)$ Carry the $(5x^3)$ term throughout the problem.

         $\dfrac{dy}{dx} = 4 \bullet 15x^2 \sec^2(5x^3)$ Multiply $4 \sec^2(5x^3)$ by the derivative of $5x^3$.

         $\dfrac{dy}{dx} = 60x^2 \sec^2(5x^3)$ Rearrange the terms and simplify.

TEACHER CERTIFICATION EXAM

SKILL 50.3   **Find the derivatives of exponential functions.**

$f(x) = e^x$ is an exponential function. The derivative of e^x is exactly the same thing $\rightarrow e^x$. If instead of x, the exponent on e is an expression, the derivative is the same e raised to the algebraic exponent multiplied by the derivative of the algebraic expression.

If a base other than e is used, the derivative is the natural log of the base times the original exponential function times the derivative of the exponent.

Examples:

1. $y = e^x$

      $\dfrac{dy}{dx} = e^x$

2. $y = e^{3x}$

      $\dfrac{dy}{dx} = e^{3x} \bullet 3$ Multiply e^{3x} by the derivative of $3x$ which is 3.

      $\dfrac{dy}{dx} = 3e^{3x}$ Rearrange the terms.

3. $y = \dfrac{5}{e^{\sin x}}$

      $y = 5e^{-\sin x}$ Rewrite using negative exponents

      $\dfrac{dy}{dx} = 5e^{-\sin x} \bullet (^-\cos x)$ Multiply $5e^{-\sin x}$ by the derivative of $^-\sin x$ which is $^-\cos x$.

      $\dfrac{dy}{dx} = \dfrac{^-5\cos x}{e^{\sin x}}$ Use the definition of negative exponents to simplify.

4. $y = {}^-2 \bullet \ln 3^{4x}$

      $\dfrac{dy}{dx} = {}^-2 \bullet (\ln 3)(3^{4x})(4)$ The natural log of the base is $\ln 3$. The derivative of $4x$ is 4.

      $\dfrac{dy}{dx} = {}^-8 \bullet 3^{4x} \ln 3$ Rearrange terms to simplify.

TEACHER CERTIFICATION EXAM

SKILL 50.4 Find the derivatives of logarithmic functions.

The most common logarithmic function on the Exam is the natural logarithmic function ($\ln x$). The derivative of $\ln x$ is simply $1/x$. If x is replaced by an algebraic expression, the derivative is the fraction one divided by the expression multiplied by the derivative of the expression.

For all other logarithmic functions, the derivative is 1 over the argument of the logarithm multiplied by 1 over the natural logarithm (ln) of the base multiplied by the derivative of the argument.

Examples:

1. $y = \ln x$

   $\dfrac{dy}{dx} = \dfrac{1}{x}$

2. $y = 3\ln(x^{-2})$

   $\dfrac{dy}{dx} = 3 \bullet \dfrac{1}{x^{-2}} \bullet \left(-2x^{-2-1}\right)$ Multiply one over the argument (x^{-2}) by the derivative of x^{-2} which is $-2x^{-2-1}$.

   $\dfrac{dy}{dx} = 3 \bullet x^2 \bullet \left(-2x^{-3}\right)$

   $\dfrac{dy}{dx} = \dfrac{-6x^2}{x^3}$ Simplify using the definition of negative exponents.

   $\dfrac{dy}{dx} = \dfrac{-6}{x}$ Cancel common factors to simplify.

3. $y = \log_5(\tan x)$

   $\dfrac{dy}{dx} = \dfrac{1}{\tan x} \bullet \dfrac{1}{\ln 5} \bullet (\sec^2 x)$ The derivative of $\tan x$ is $\sec^2 x$.

   $\dfrac{dy}{dx} = \dfrac{\sec^2 x}{(\tan x)(\ln 5)}$

MATHEMATICS HIGH SCHOOL

SKILL 50.5 Find the derivative of the implicitly defined function of x, using implicit differentiation.

Implicitly defined functions are ones where both variables (usually x and y) appear in the function. All of the rules for finding the derivative still apply to both variables. The only difference is that, while the derivative of x is one (1) and typically not even mentioned, the derivative of y must be written y' or dy/dx. Work these problems just like the other derivative problems, just remember to include the extra step of multiplying by dy/dx in the result.

If the question asks for the derivative of y, given an equation with x and y on both sides, find the derivative of each side. Then solve the new equation for dy/dx just as you would an algebra problem.

Examples:

1. $\dfrac{d}{dx}(y^3) = 3y^{3-1} \cdot \dfrac{dy}{dx}$ Recall the derivative of x^3 is

   $\dfrac{d}{dx}(y^3) = 3y^2 \dfrac{dy}{dx}$ $3x^{3-1}$. Follow the same rule, but also multiply by the derivative of y which is dy/dx.

2. $\dfrac{d}{dx}(3\ln y) = 3 \cdot \dfrac{1}{y} \cdot \dfrac{dy}{dx}$

3. $\dfrac{d}{dx}(^-2\cos y) = {}^-2(^-1\sin y)\dfrac{dy}{dx}$ Recall the derivative of $\cos x$ is

   $\dfrac{d}{dx}(^-2\cos y) = 2\sin y \dfrac{dy}{dx}$ $^-\sin x$.

4. $2y = e^{3x}$ Solve for after taking the derivative.

   $2 \cdot \dfrac{dy}{dx} = e^{3x} \cdot 3$ The derivative of e^{3x} is $e^{3x} \cdot 3$.

   $\dfrac{dy}{dx} = \dfrac{3}{2}e^{3x}$ Divide both sides by 2 to solve for the derivative dy/dx.

SKILL 50.6 Find the derivative of the sum, product, or quotient of functions.

A. Derivative of a sum--find the derivative of each term separately and add the results.

B. Derivative of a product--multiply the derivative of the first factor by the second factor and add to it the product of the first factor and the derivative of the second factor.

Remember the phrase "first times the derivative of the second plus the second times the derivative of the first."

C. Derivative of a quotient--use the rule "bottom times the derivative of the top minus the top times the derivative of the bottom all divided by the bottom squared."

Examples:

1. $y = 3x^2 + 2\ln x + 5\sqrt{x}$ $\qquad \sqrt{x} = x^{1/2}$.

   $\dfrac{dy}{dx} = 6x^{2-1} + 2 \cdot \dfrac{1}{x} + 5 \cdot \dfrac{1}{2} x^{1/2-1}$

   $\dfrac{dy}{dx} = 6x + \dfrac{2}{x} + \dfrac{5}{2} \cdot \dfrac{1}{\sqrt{x}} = \dfrac{12x^2 + 4x + 5\sqrt{x}}{2x}$ $\qquad x^{1/2-1} = x^{-1/2} = 1/\sqrt{x}$.

2. $y = 4e^{x^2} \bullet \sin x$

   $\dfrac{dy}{dx} = 4(e^{x^2} \bullet \cos x + \sin x \bullet e^{x^2} \bullet 2x)$ $\qquad$ The derivative of

   e^{x^2} is $e^{x^2} \bullet 2$.

   $\dfrac{dy}{dx} = 4(e^{x^2} \cos x + 2xe^{x^2} \sin x)$

   $\dfrac{dy}{dx} = 4e^{x^2} \cos x + 8xe^{x^2} \sin x$

3. $y = \dfrac{\cos x}{x}$

   $\dfrac{dy}{dx} = \dfrac{x(-\sin x) - \cos x \bullet 1}{x^2}$ $\qquad$ The derivative of x is 1.

   $\qquad\qquad\qquad\qquad\qquad\qquad\qquad$ The derivative of $\cos x$ is $-\sin x$.

   $\dfrac{dy}{dx} = \dfrac{-x \sin x - \cos x}{x^2}$

SKILL 50.7 Find the derivatives of a composite function (chain rule).

A composite function is made up of two or more separate functions such as $\sin(\ln x)$ or $x^2 e^{3x}$. To find the derivatives of these composite functions requires two steps. First identify the predominant function in the problem. For example, in $\sin(\ln x)$ the major function is the sine function. In $x^2 e^{3x}$ the major function is a product of two expressions (x^2 and e^{3x}). Once the predominant function is identified, apply the appropriate differentiation rule. Be certain to include the step of taking the derivative of every part of the functions which comprise the composite function. Use parentheses as much as possible.

Examples:

1. $y = \sin(\ln x)$ — The major function is a sine function.

    $\dfrac{dy}{dx} = [\cos(\ln x)] \bullet \left[\dfrac{1}{x}\right]$ — The derivative of $\sin x$ is $\cos x$. The derivative of $\ln x$ is $1/x$.

2. $y = x^2 \bullet e^{3x}$ — The major function is a product.

    $\dfrac{dy}{dx} = x^2 \left(e^{3x} \bullet 3\right) + e^{3x} \bullet 2x$ — The derivative of a product is "First times the derivative of second plus the second times the derivative of the first."

    $\dfrac{dy}{dx} = 3x^2 e^{3x} + 2x e^{3x}$

3. $y = \tan^2\left(\dfrac{\ln x}{\cos x}\right)$ — This function is made of several functions. The major function is a power function.

    $\dfrac{dy}{dx} = \left[2\tan^{2-1}\left(\dfrac{\ln x}{\cos x}\right)\right]\left[\sec^2\left(\dfrac{\ln x}{\cos x}\right)\right]\left[\dfrac{d}{dx}\left(\dfrac{\ln x}{\cos x}\right)\right]$

    The derivative of $\tan x$ is $\sec^2 x$. Hold off one more to take the derivative of $\ln x / \cos x$.

    $\dfrac{dy}{dx} = \left[2\tan\left(\dfrac{\ln x}{\cos x}\right)\sec^2\left(\dfrac{\ln x}{\cos x}\right)\right]\left[\dfrac{(\cos x)(1/x) - \ln x(^-\sin x)}{\cos^2 x}\right]$

    $\dfrac{dy}{dx} = \left[2\tan\left(\dfrac{\ln x}{\cos x}\right)\sec^2\left(\dfrac{\ln x}{\cos x}\right)\right]\left[\dfrac{(\cos x)(1/x) + \ln x(\sin x)}{\cos^2 x}\right]$

    The derivative of a quotient is "Bottom times the derivative of the top minus the top times the derivative of the bottom all divided by the bottom squared."

SKILL 50.8 Find higher order derivatives.

If a question simply asks for the derivative of a function, the question is asking for the first derivative. To find the second derivative of a function, take the derivative of the first derivative. To find the third derivative, take the derivative of the second derivative; and so on. All of the regular derivative rules still apply.

Examples:

1. Find the second derivative $\left(\dfrac{d^2y}{dx^2} \text{ or } y''\right)$ of the following function:

$$y = 5x^2$$

$\dfrac{dy}{dx} = 2 \bullet 5x^{2-1} = 10x$ Take the first derivative.

$\dfrac{d^2y}{dx^2} = 10$ The derivative of $10x$ is 10.

2. Find the third derivative (y''') of the following function:

$$y = e^{\sin x}$$

$y' = e^{\sin x} \bullet \cos x$ Take the first derivative. Use the chain rule. The derivative of $\sin x$ is $\cos x$.

$y'' = \left(e^{\sin x}\right)\left(-\sin x\right) + \left(\cos x\right)\left(e^{\sin x} \cos x\right)$

The major function here is a product. Use the product rule to take the derivative again. Remember, "First times the derivative of second plus the second times the derivative of the first."

$y''' = -(\sin x)e^{\sin x} + \left(\cos^2 x\right)e^{\sin x}$ Simplify.

SKILL 50.9 Apply Rolle's theorem and the mean value theorem.

Rolle's Theorem--if there are two values for x, say 2 and -3, where $f(2) = f(^-3)$, then there is some number between $^-3$ and 2 where the derivative at that number equals zero. In other words, there is a point somewhere between $x = ^-3$ and $x = 2$ where the graph reaches a maximum or a minimum, like in a parabola.

Mean Value Theorem--pick two points on any curve, say $(a, f(a))$ and $(b, f(b))$. If you draw a line between these two points, the slope of it would be the change in y over the change in x or
$$slope = \frac{f(b) - f(a)}{b - a}.$$
The Mean Value Theorem says that there will be some point on the curve between $x = a$ and $x = b$ where the derivative is also equal to this slope. Set the derivative equal to this slope value and solve; doing so will produce the point where the tangent line is parallel to the line formed by connecting $(a, f(a))$ and $(b, f(b))$. Try to use the Mean Value Theorem if the problem refers to *average* rates of change.

Example:
1. the position of a particle is given by the equation $s(t) = \sin t$. What is does Rolle's Theorem tell you about motion of the particle during the interval $(0, 2\pi)$?

To use Rolle's Theorem you must first verify that $\sin t$ equals the same number for $t = 0$ and $t = 2\pi$. ($\sin 0 = 0$ and $\sin 2\pi = 0$)

The graph of $s(t)$ shows these 2 points: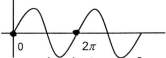

So, Rolle's Theorem states that there is at least a value between 0 and 2π so that the derivative of $s(t)$ is equal to zero.
$$\frac{ds}{dx} = \cos t$$
Set the derivative equal to zero and solve.

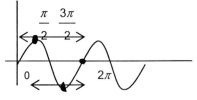

$\cos(t) = 0$

$t = \frac{\pi}{2}$, and $\frac{3\pi}{2}$

So, at $t = \frac{\pi}{2}$ and $t = \frac{3\pi}{2}$ the graph makes a turn and the tangent line is horizontal.

SKILL 50.10 Recognize and apply definitions of the derivative of a function.

The derivative of a function has two basic interpretations.

    I.   Instantaneous rate of change
    II.   Slope of a tangent line at a given point

If a question asks for the rate of change of a function, take the derivative to find the equation for the rate of change. Then plug in for the variable to find the instantaneous rate of change.

The following is a list summarizing some of the more common quantities referred to in rate of change problems.

area	height	profit
decay	population growth	sales
distance	position	temperature
frequency	pressure	volume

Pick a point, say $x = {^-}3$, on the graph of a function. Draw a tangent line at that point. Find the derivative of the function and plug in $x = {^-}3$. The result will be the slope of the tangent line.

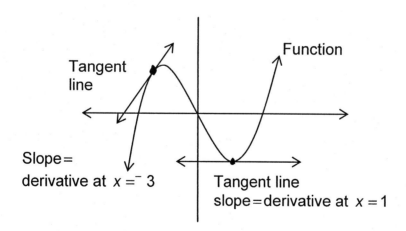

COMPETENCY 51.0   **ABILITY TO APPLY DERIVATIVES TO FIND THE SLOPES OF CURVES AND TANGENTS AND NORMAL LINES TO A CURVE.**

SKILL 51.1   **Use the derivative to find the slope of a curve at a point.**

To find the slope of a curve at a point, there are two steps to follow.

1. Take the derivative of the function.
2. Plug in the value to find the slope.

If plugging into the derivative yields a value of zero, the tangent line is horizontal at that point.

If plugging into the derivative produces a fraction with zero in the denominator, the tangent line at this point has an undefined slope and is thus a vertical line.

Examples:

1. Find the slope of the tangent line for the given function at the given point.

$$y = \frac{1}{x-2} \text{ at } (3,1)$$

$y = (x-2)^{-1}$ Rewrite using negative exponents.

$\dfrac{dy}{dx} = {}^-1(x-2)^{-1-1}(1)$ Use the Chain rule. The derivative of $(x-2)$ is 1.

$\dfrac{dy}{dx} = {}^-1(x-2)^{-2}$

$\dfrac{dy}{dx}\bigg|_{x=3} = {}^-1(3-2)^{-2}$ Evaluate at $x=3$.

$\dfrac{dy}{dx}\bigg|_{x=3} = {}^-1$ The slope of the tangent line is $^-1$ at $x=3$.

2. Find the points where the tangent to the curve $f(x) = 2x^2 + 3x$ is parallel to the line $y = 11x - 5$.

$f'(x) = 2 \bullet 2x^{2-1} + 3$ Take the derivative of $f(x)$ to get the slope of a tangent line.

$f'(x) = 4x + 3$

$4x + 3 = 11$ Set the slope expression $(4x + 3)$ equal to the slope of $y = 11x - 5$.

$x = 2$ Solve for the x value of the point.

$f(2) = 2(2)^2 + 3(2)$ The y value is 14.

$f(2) = 14$ So $(2, 14)$ is the point on $f(x)$ where the tangent line is parallel to $y = 11x - 5$.

SKILL 51.2 Find the equation of a tangent line at a point on the curve.

To write an equation of a tangent line at a point, two things are needed.

A point--the problem will usually provide a point, (x, y). If the problem only gives an x value, plug the value into the original function to get the y coordinate.

The slope--to find the slope, take the derivative of the original function. Then plug in the x value of the point to get the slope.

After obtaining a point and a slope, use the Point-Slope form for the equation of a line:

$$(y - y_1) = m(x - x_1)$$

where m is the slope and (x_1, y_1) is the point.

Example:

Find the equation of the tangent line to $f(x) = 2e^{x^2}$ at $x = {}^-1$.

$f({}^-1) = 2e^{({}^-1)^2}$	Plug in the x coordinate to obtain the y coordinate.
$= 2e^1$	The point is $({}^-1, 2e)$.
$f'(x) = 2e^{x^2} \bullet (2x)$	
$f'({}^-1) = 2e^{({}^-1)^2} \bullet (2 \bullet {}^-1)$	
$f'({}^-1) = 2e^1({}^-2)$	
$f'({}^-1) = {}^-4e$	The slope at $x = {}^-1$ is ${}^-4e$.
$(y - 2e) = {}^-4e(x - {}^-1)$	Plug in the point $({}^-1, 2e)$ and the slope $m = {}^-4e$. Use the point slope form of a line.
$y = {}^-4ex - 4e + 2e$	
$y = {}^-4ex - 2e$	Simplify to obtain the equation for the tangent line.

SKILL 51.3 Find the equation of a normal line at a point on the curve.

A normal line is a line which is perpendicular to a tangent line at a given point. Perpendicular lines have slopes which are negative reciprocals of each other. To find the equation of a normal line, first get the slope of the tangent line at the point (see 51.2). Find the negative reciprocal of this slope. Next, use the new slope and the point on the curve, both the x_1 and y_1 coordinates, and substitute into the Point-Slope form of the equation for a line:

$$(y - y_1) = slope \bullet (x - x_1)$$

Examples:

1. Find the equation of the normal line to the tangent to the curve $y = (x^2 - 1)(x - 3)$ at $x = {}^-2$.

$f(^-2) = ((^-2)^2 - 1)(^-2 - 3)$	First find the y coordinate of the point on the curve. Here,
$f(^-2) = {}^-15$	$y = {}^-15$ when $x = {}^-2$.
$y = x^3 - 3x^2 - x + 3$	Before taking the derivative, multiply the expression first. The derivative of sum is easier to find than the derivative of a product.
$y' = 3x^2 - 6x - 1$	Take the derivative to find the slope of the tangent line.
$y'_{x={}^-2} = 3(^-2)^2 - 6(^-2) - 1$	
$y'_{x={}^-2} = 23$	
slope of normal $= \dfrac{{}^-1}{23}$	For the slope of the normal line, take the negative reciprocal of the tangent line's slope.
$(y - {}^-15) = \dfrac{{}^-1}{23}(x - {}^-2)$	Plug (x_1, y_1) into the point-slope
$(y + 15) = \dfrac{{}^-1}{23}(x + 2)$	equation.

2. Find the equation of the normal line to the tangent to the curve $y = \ln(\sin x)$ at $x = \pi$.

$f(\pi) = \ln(\sin \pi)$	$\sin \pi = 1$ and $\ln(1) = 0$ (recall $e^0 = 1$).
$f(\pi) = \ln(1) = 0$	So $x_1 = \pi$ and $y_1 = 0$.
$y' = \dfrac{1}{\sin x} \bullet \cos x$	Take the derivative to find the slope of the tangent line.
$y'_{x=\pi} = \dfrac{\cos \pi}{\sin \pi} = \dfrac{0}{1}$	
Slope of normal does not exist.	$\dfrac{{}^-1}{0}$ does not exist. So the normal Line is vertical at $x = \pi$.

MATHEMATICS HIGH SCHOOL

COMPETENCY 52.0   **ABILITY TO IDENTIFY INCREASING AND DECREASING FUNCTIONS, RELATIVE AND ABSOLUTE MAXIMUM AND MINIMUM POINTS, CONCAVITY, AND POINTS OF INFLECTION.**

SKILL 52.1   **Determine if a function is increasing or decreasing by using the first derivative in a given interval.**

A function is said to be increasing if it is rising from left to right and decreasing if it is falling from left to right. Lines with positive slopes are increasing, and lines with negative slopes are decreasing. If the function in question is something other than a line, simply refer to the slopes of the tangent lines as the test for increasing or decreasing. Take the derivative of the function and plug in an x value to get the slope of the tangent line; a positive slope means the function is increasing and a negative slope means it is decreasing. If an interval for x values is given, just pick any point between the two values to substitute.

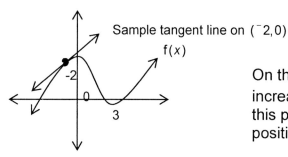

Sample tangent line on $(^-2, 0)$

On the interval $(^-2, 0)$, $f(x)$ is increasing. The tangent lines on this part of the graph have positive slopes.

Example:

The growth of a certain bacteria is given by $f(x) = x + \dfrac{1}{x}$. Determine if the rate of growth is increasing or decreasing on the time interval $(^-1, 0)$.

$$f'(x) = 1 + \dfrac{^-1}{x^2}$$

To test for increasing or decreasing, find the slope of the tangent line by taking the derivative.

$$f'\left(\dfrac{^-1}{2}\right) = 1 + \dfrac{^-1}{(^-1/2)^2}$$

Pick any point on $(^-1, 0)$ and substitute into the derivative.

$$f'\left(\dfrac{^-1}{2}\right) = 1 + \dfrac{^-1}{1/4}$$

The slope of the tangent line at $x = \dfrac{^-1}{2}$

$$= 1 - 4$$
$$= ^-3$$

is $^-3$. The exact value of the slope is not important. The important fact is that the slope is negative.

SKILL 52.2 Find relative and absolute maxima and minima.

Substituting an x value into a function produces a corresponding y value. The coordinates of the point (x,y), where y is the largest of all the y values, is said to be a maximum point. The coordinates of the point (x,y), where y is the smallest of all the y values, is said to be a minimum point. To find these points, only a few x values must be tested. First, find all of the x values that make the derivative either zero or undefined. Substitute these values into the original function to obtain the corresponding y values. Compare the y values. The largest y value is a maximum; the smallest y value is a minimum. If the question asks for the maxima or minima on an interval, be certain to also find the y values that correspond to the numbers at either end of the interval.

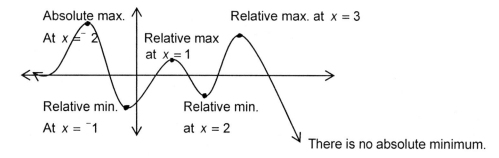

Example:

Find the maxima and minima of $f(x) = 2x^4 - 4x^2$ at the interval $(^-2, 1)$.

$f'(x) = 8x^3 - 8x$ Take the derivative first. Find all the
$8x^3 - 8x = 0$ x values (critical values) that make the
$8x(x^2 - 1) = 0$ derivative zero or undefined. In this
$8x(x-1)(x+1) = 0$ case, there are no x values that make
$x = 0, x = 1, $ or $x = ^-1$ the derivative undefined.
$f(0) = 2(0)^4 - 4(0)^2 = 0$ Substitute the critical values into the
$f(1) = 2(1)^4 - 4(1)^2 = ^-2$ original function. Also, plug in the
$f(^-1) = 2(^-1)^4 - 4(^-1)^2 = ^-2$ endpoint of the interval. Note that 1 is
$f(^-2) = 2(^-2)^4 - 4(^-2)^2 = 16$ a critical point and an endpoint.

The maximum is at $(^-2, 16)$ and there are minima at $(1, ^-2)$ and $(^-1, ^-2)$. $(0,0)$ is neither the maximum or minimum on $(^-2, 1)$ but it is still considered a relative extra point.

SKILL 52.3 Find intervals on a curve where the curve is concave up or concave down.

The first derivative reveals whether a curve is rising or falling (increasing or decreasing) from the left to the right. In much the same way, the second derivative relates whether the curve is concave up or concave down. Curves which are concave up are said to "collect water;" curves which are concave down are said to "dump water." To find the intervals where a curve is concave up or concave down, follow the following steps.

1. Take the second derivative (i.e. the derivative of the first derivative).
2. Find the critical x values.
   - Set the second derivative equal to zero and solve for critical x values.
   - Find the x values that make the second derivative undefined (i.e. make the denominator of the second derivative equal to zero). Such values may not always exist.
3. Pick sample values which are both less than and greater than each of the critical values.
4. Substitute each of these sample values into the second derivative and determine whether the result is positive or negative.
   - If the sample value yields a positive number for the second derivative, the curve is concave up on the interval where the sample value originated.
   - If the sample value yields a negative number for the second derivative, the curve is concave down on the interval where the sample value originated.

Example:

Find the intervals where the curve is concave up and concave down for $f(x) = x^4 - 4x^3 + 16x - 16$.

$f'(x) = 4x^3 - 12x^2 + 16$ — Take the second derivative.

$f''(x) = 12x^2 - 24x$ — Find the critical values by setting the

$12x^2 - 24x = 0$ — second derivative equal to zero.

$12x(x - 2) = 0$ — There are no values that make the

$x = 0$ or $x = 2$ — second derivative undefined.

Set up a number line with the critical values.

Sample values: $^-1, 1, 3$ — Pick sample values in each of the 3

$f''(^-1) = 12(^-1)^2 - 24(^-1) = 36$ — intervals. If the sample value

$f''(1) = 12(1)^2 - 24(1) = ^-12$ — produces a negative number,

$f''(3) = 12(3)^2 - 24(3) = 36$ — the function is concave down.

TEACHER CERTIFICATION EXAM

If the value produces a positive number, the curve is concave up. If The value produces a zero, the function is linear.

Therefore when $x < 0$ the function is concave up,
when $0 < x < 2$ the function is concave down,
when $x > 2$ the function is concave up.

SKILL 52.4 Identify points of inflection.

A point of inflection is a point where a curve changes from being concave up to concave down or vice versa. To find these points, follow the steps in section 52.3 for finding the intervals where a curve is concave up or concave down. A critical value is part of an inflection point if the curve is concave up on one side of the value and concave down on the other. The critical value is the x coordinate of the inflection point. To get the y coordinate, plug the critical value into the **original** function.

Example: Find the inflection points of $f(x) = 2x - \tan x$ where $\frac{-\pi}{2} < x < \frac{\pi}{2}$.

$(x) = 2x - \tan x \quad \frac{-\pi}{2} < x < \frac{\pi}{2}$	Note the restriction on x.
$f'(x) = 2 - \sec^2 x$	Take the second derivative.
$f''(x) = 0 - 2 \cdot \sec x \cdot (\sec x \tan x)$	Use the Power rule.
$= {}^-2 \cdot \dfrac{1}{\cos x} \cdot \dfrac{1}{\cos x} \cdot \dfrac{\sin x}{\cos x}$	The derivative of $\sec x$ is $(\sec x \tan x)$.
$f''(x) = \dfrac{{}^-2\sin x}{\cos^3 x}$	Find critical values by solving for the second derivative equal to zero.
$0 = \dfrac{{}^-2\sin x}{\cos^3 x}$	No x values on $\left(\dfrac{-\pi}{2}, \dfrac{\pi}{2}\right)$ make the denominator zero.
${}^-2\sin x = 0$	
$\sin x = 0$	Pick sample values on each side of the critical value $x = 0$.
$x = 0$	

Sample values: $x = \dfrac{-\pi}{4}$ and $x = \dfrac{\pi}{4}$

MATHEMATICS HIGH SCHOOL

$$f''\left(\frac{-\pi}{4}\right) = \frac{-2\sin(-\pi/4)}{\cos^3(\pi/4)} = \frac{-2(-\sqrt{2}/2)}{(\sqrt{2}/2)^3} = \frac{\sqrt{2}}{(\sqrt{8}/8)} = \frac{8\sqrt{2}}{\sqrt{8}}$$

$$f''\left(\frac{\pi}{4}\right) = \frac{-2\sin(\pi/4)}{\cos^3(\pi/4)} = \frac{-2(\sqrt{2}/2)}{(\sqrt{2}/2)^3} = \frac{-\sqrt{2}}{(\sqrt{8}/8)} = \frac{-8\sqrt{2}}{\sqrt{8}}$$

The second derivative is positive on $(0,\infty)$ and negative on $(-\infty,0)$. So the curve changes concavity at $x = 0$. Use the original equation to find the y value that inflection occurs at.

$f(0) = 2(0) - \tan 0 = 0 - 0 = 0$ The inflection point is (0,0).

SKILL 52.5 Solve extreme value problems.

Extreme value problems are also known as max-min problems. Extreme value problems require using the first derivative to find values which either maximize or minimize some quantity such as area, profit, or volume. Follow these steps to solve an extreme value problem.

1. Write an equation for the quantity to be maximized or minimized.
2. Use the other information in the problem to write secondary equations.
3. Use the secondary equations for substitutions, and rewrite the original equation in terms of only one variable.
4. Find the derivative of the primary equation (step 1) and the critical values of this derivative.
5. Substitute these critical values into the primary equation. The value which produces either the largest or smallest value is used to find the solution.

Example:

A manufacturer wishes to construct an open box from the piece of metal shown below by cutting squares from each corner and folding up the sides. The square piece of metal is 12 feet on a side. What are the dimensions of the squares to be cut out which will maximize the volume?

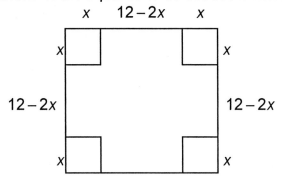

Volume = lwh	Primary equation.
$l = 12 - 2x$	
$w = 12 - 2x$	Secondary equations.
$h = x$	
$V = (12 - 2x)(12 - 2x)(x)$	Make substitutions.
$V = (144x - 48x^2 + 4x^3)$	Take the derivative.
$\dfrac{dV}{dx} = 144 - 96x + 12x^2$	Find critical values by setting the
$0 = 12(x^2 - 8x + 12)$	derivative equal to zero.
$0 = 12(x - 6)(x - 2)$	
$x = 6$ and $x = 2$	Substitute critical values into volume equation.
$V = 144(6) - 48(6)^2 + 4(6)^3$ $V = 144(2) - 48(2)^2 + 4(2)^3$	
$V = 0$ ft^3 when $x = 4$ $V = 128$ ft^3 when $x = 2$	

Therefore, the manufacturer can maximize the volume if the squares to be cut out are 2 feet by 2 feet ($x = 2$).

SKILL 52.6 Solve problems using velocity and acceleration of a particle moving along a line.

If a particle (or a car, a bullet, etc.) is moving along a line, then the distance that the particle travels can be expressed by a function in terms of time.

1. The first derivative of the distance function will provide the velocity function for the particle. Substituting a value for time into this expression will provide the instantaneous velocity of the particle at the time. Velocity is the rate of change of the distance traveled by the particle. Taking the absolute value of the derivative provides the speed of the particle. A positive value for the velocity indicates that the particle is moving forward, and a negative value indicates the particle is moving backwards.
2. The second derivative of the distance function (which would also be the first derivative of the velocity function) provides the acceleration function. The acceleration of the particle is the rate of change of the velocity. If a value for time produces a positive acceleration, the particle is speeding up; if it produces a negative value, the particle is slowing down. If the acceleration is zero, the particle is moving at a constant speed.

To find the time when a particle stops, set the first derivative (i.e. the velocity function) equal to zero and solve for time. This time value is also the instant when the particle changes direction.

Example:

The motion of a particle moving along a line is according to the equation: $s(t) = 20 + 3t - 5t^2$ where s is in meters and t is in time. Find the position, velocity, and acceleration of a particle at $t = 2$ seconds.

$s(2) = 20 + 3(2) - 5(2)^2$ $= 6$ meters	Plug $t = 2$ into the original equation to find the position.
$s'(t) = v(t) = 3 - 10t$	The derivative of the first function gives the velocity.
$v(2) = 3 - 10(2) = {}^-17$ m/s	Plug $t = 2$ into the velocity function to find the velocity. ${}^-17$ m/s indicates the particle is moving backwards.
$s''(t) = a(t) = {}^-10$ $a(2) = {}^-10$ m/s	The second derivation of position gives the acceleration. Substitute $t = 2$, yields an acceleration of ${}^-10$ m/s, which indicates the particle is slowing down.

SKILL 52.7 Solve problems using instantaneous rates of change and related rates of change.

Finding the rate of change of one quantity (for example distance, volume, etc.) with respect to time it is often referred to as a rate of change problem. To find an instantaneous rate of change of a particular quantity, write a function in terms of time for that quantity; then take the derivative of the function. Substitute in the values at which the instantaneous rate of change is sought.

Functions which are in terms of more than one variable may be used to find related rates of change. These functions are often not written in terms of time. To find a related rate of change, follow these steps.

1. Write an equation which relates all the quantities referred to in the problem.
2. Take the derivative of both sides of the equation with respect to time. Follow the same steps as used in implicit differentiation (Section 50.5). This means take the derivative of each part of the equation remembering to multiply each term by the derivative of the variable involved with respect to time. For example, if a term includes the variable v for volume, take the derivative of the term remembering to multiply by dv/dt for the derivative of volume with respect to time. dv/dt is the rate of change of the volume.
3. Substitute the known rates of change and quantities, and solve for the desired rate of change.

Example:

1. What is the instantaneous rate of change of the area of a circle where the radius is 3 cm?

$A(r) = \pi r^2$	Write an equation for area.
$A'(r) = 2\pi r$	Take the derivative to find the rate of change.
$A'(3) = 2\pi(3) = 6\pi$	Substitute in $r = 3$ to arrive at the instantaneous rate of change.

TEACHER CERTIFICATION EXAM

COMPETENCY 53.0   ABILITY TO FIND ANTIDERIVATIVES.

SKILL 53.1   Find antiderivatives for algebraic functions.

Taking the antiderivative of a function is the opposite of taking the derivative of the function--much in the same way that squaring an expression is the opposite of taking the square root of the expression. For example, since the derivative of x^2 is $2x$ then the antiderivative of $2x$ is x^2. The key to being able to take antiderivatives is being as familiar as possible with the derivative rules.

To take the antiderivative of an algebraic function (the sum of products of coefficients and variables raised to powers other than negative one), take the antiderivative of each term in the function by following these steps.

1. Take the antiderivative of each term separately.
2. The coefficient of the variable should be equal to one plus the exponent.
3. If the coefficient is not one more than the exponent, put the correct coefficient on the variable and also multiply by the reciprocal of the number put in.
    Ex. For $4x^5$, the coefficient should be 6 not 4. So put in 6 and the reciprocal 1/6 to achieve $(4/6)6x^5$.
4. Finally take the antiderivative by replacing the coefficient and variable with just the variable raised to one plus the original exponent.
    Ex. For $(4/6)6x^5$, the antiderivative is $(4/6)x^6$.
    You have to add in constant c because there is no way to know if a constant was originally present since the derivative of a constant is zero.
5. Check your work by taking the first derivative of your answer. You should get the original algebraic function.

Examples: Take the antiderivative of each function.

1. $f(x) = 5x^4 + 2x$ The coefficient of each term is already 1 more than the exponent.
   $F(x) = x^5 + x^2 + c$ $F(x)$ is the antiderivative of $f(x)$
   $F'(x) = 5x^4 + 2x$ Check by taking the derivative of $F(x)$.

2. $f(x) = {}^-2x^{-3}$
   $F(x) = x^{-2} + c = \dfrac{1}{x^2} + c$ $F(x)$ is the antiderivative of $f(x)$.
   $F'(x) = {}^-2x^{-3}$ Check.

MATHEMATICS HIGH SCHOOL

3. $f(x) = {}^-4x^2 + 2x^7$ Neither coefficient is correct.

$f(x) = {}^-4 \cdot \dfrac{1}{3} \cdot 3x^2 + 2 \cdot \dfrac{1}{8} \cdot 8x^7$ Put in the correct coeffient along with its reciprocal.

$F(x) = {}^-4 \cdot \dfrac{1}{3} x^3 + 2 \cdot \dfrac{1}{8} x^8 + c$ $F(x)$ is the antiderivative of $f(x)$.

$F(x) = \dfrac{{}^-4}{3} x^3 + \dfrac{1}{4} x^8 + c$

$F'(x) = {}^- 4x^2 + 2x^7$ Check.

TEACHER CERTIFICATION EXAM

SKILL 53.2 Find antiderivatives for trigonometric functions.

The rules for taking antiderivatives of trigonometric functions follow, but be aware that these can get very confusing to memorize because they are very similar to the derivative rules. Check the antiderivative you get by taking the derivative and comparing it to the original function.

1. $\sin x$ the antiderivative for $\sin x$ is $^-\cos x + c$.
2. $\cos x$ the antiderivative for $\cos x$ is $\sin x + c$.
3. $\tan x$ the antiderivative for $\tan x$ is $\ln|\cos x| + c$.
4. $\sec^2 x$ the antiderivative for $\sec^2 x$ is $\tan x + c$.
5. $\sec x \tan x$ the antiderivative for $\sec x \tan x$ is $\sec x + c$.

If the trigonometric function has a coefficient, simply keep the coefficient and multiply the antiderivative by it.

Examples: Find the antiderivatives for the following functions.

1. $f(x) = 2\sin x$ Carry the 2 throughout the problem.
$F(x) = 2(^-\cos x) = {}^-2\cos x + c$ $F(x)$ is the antiderivative.
$F'(x) = {}^-2(^-\sin x) = 2\sin x$ Check by taking the derivative of $F(x)$.

2. $f(x) = \dfrac{\tan x}{5}$

$F(x) = \dfrac{\ln|\cos x|}{5} + c$ $F(x)$ is the antiderivative of $f(x)$.

$F'(x) = \dfrac{1}{5}\left(\dfrac{1}{|\cos x|}\right)(\sin x) = \dfrac{1}{5}\tan x$ Check by taking the derivative of $F(x)$.

Practice problems: Find the antiderivative of each function.

1. $f(x) = {}^-20\cos x$ 2. $f(x) = \pi \sec x \tan x$

SKILL 53.3 Find antiderivatives for exponential functions.

Use the following rules when finding the antiderivative of an exponential function.

1. e^x The antiderivative of e^x is the same $e^x + c$.
2. a^x The antiderivative of a^x, where a is any number, is $a^x / \ln a + c$.

Examples: Find the antiderivatives of the following functions:

1. $f(x) = 10e^x$
   $F(x) = 10e^x + c$ $F(x)$ is the antiderivative.
   $F'(x) = 10e^x$ Check by taking the derivative of $F(x)$.

2. $f(x) = \dfrac{2^x}{3}$
   $F(x) = \dfrac{1}{3} \cdot \dfrac{2^x}{\ln 2} + c$ $F(x)$ is the antiderivative.
   $F'(x) = \dfrac{1}{3\ln 2} \ln 2 (2^x)$ Check by taking the derivative of $F(x)$.
   $F'(x) = \dfrac{2^x}{3}$

COMPETENCY 54.0   **UNDERSTANDING OF ANTIDERIVATIVE PROBLEMS RELATED TO MOTIONS OF BODIES.**

SKILL 54.1   **Solve velocity problems.**

The derivative of a distance function provides a velocity function, and the derivative of a velocity function provides an acceleration function. Therefore taking the antiderivative of an acceleration function yields the velocity function, and the antiderivative of the velocity function yields the distance function.

Example:

A particle moves along the x axis with acceleration $a(t) = 3t - 1$ cm/sec/sec. At time $t = 4$, the particle is moving to the left at 3 cm per second. Find the velocity of the particle at time $t = 2$ seconds.

$a(t) = 3t - 1$ Before taking the antiderivative, make

$a(t) = 3 \cdot \dfrac{1}{2} \cdot 2t - 1$ sure the correct coefficients are present.

$v(t) = \dfrac{3}{2}t^2 - 1 \cdot t + c$ $v(t)$ is the antiderivative of $a(t)$.

$v(4) = \dfrac{3}{2}(4)^2 - 1(4) + c = {}^-3$ Use the given information that $v(4) = {}^-3$

$24 - 4 + c = {}^-3$ to find c.
$20 + c = {}^-3$
$c = {}^-23$ The constant is ${}^-23$.

$v(t) = \dfrac{3}{2}t^2 - 1t + {}^-23$ Rewrite $v(t)$ using $c = {}^-23$.

$v(2) = \dfrac{3}{2}2^2 + 1(2) + {}^-23$ Solve $v(t)$ at $t=2$.

$v(2) = 6 - 2 + {}^-23 = {}^-19$ the velocity at $t = 2$ is ${}^-19$ cm/sec.

Practice problem:

A particle moves along a line with acceleration $a(t) = 5t + 2$. The velocity after 2 seconds is ${}^-10$ m/sec.

1. Find the initial velocity.
2. Find the velocity at $t = 4$.

SKILL 54.2 Solve distance problems.

To find the distance function, take the antiderivative of the velocity function. And to find the velocity function, find the antiderivative of the acceleration function. Use the information in the problem to solve for the constants that result from taking the antiderivatives.

Example:

A particle moves along the x axis with acceleration $a(t) = 6t - 6$. The initial velocity is 0 m/sec and the initial position is 8 cm to the right of the origin. Find the velocity and position functions.

$v(0) = 0$
$s(0) = 8$ — Interpret the given information.

$a(t) = 6t - 6$ — Put in the coefficients needed to take the antiderivative.
$a(t) = 6 \cdot \dfrac{1}{2} \cdot 2t - 6$

$v(t) = \dfrac{6}{2}t^2 - 6t + c$ — Take the antiderivative of $a(t)$ to get $v(t)$.

$v(0) = 3(0)^3 - 6(0) + c = 0$ — Use $v(0) = 0$ to solve for c.
$0 - 0 + c = 0$
$c = 0$
$\qquad\qquad\qquad\qquad c = 0$

$v(t) = 3t^2 - 6t + 0$ — Rewrite $v(t)$ using $c = 0$.

$v(t) = 3t^2 - 6\dfrac{1}{2} \cdot 2t$ — Put in the coefficients needed to take the antiderivative.

$s(t) = t^3 - \dfrac{6}{2}t^2 + c$ — Take the antiderivative of $v(t)$ to get $s(t)$ → the distance function.

$s(0) = 0^3 - 3(0)^2 + c = 8$ — Use $s(0) = 8$ to solve for c.
$\qquad\qquad c = 8$
$s(t) = t^3 - 3t^2 + 8$

MATHEMATICS HIGH SCHOOL

COMPETENCY 55.0   **COMPREHENSION OF TECHNIQUES OF INTEGRATION AND THEIR USE.**

SKILL 55.1   Evaluate an integral by use of the fundamental theorem of calculus.

An integral is almost the same thing as an antiderivative, the only difference is the notation.

$\int_{-2}^{1} 2x\,dx$ is the integral form of the antiderivative of $2x$. The numbers at the to top and bottom of the integral sign (1 and $^-2$) are the numbers used to find the exact value of this integral. If these numbers are used the integral is said to be *definite* and does not have an unknown constant c in the answer.

The fundamental theorem of calculus states that an integral such as the one above is equal to the antiderivative of the function inside (here $2x$) evaluated from $x = {}^-2$ to $x = 1$. To do this, follow these steps.

1. Take the antiderivative of the function inside the integral.
2. Plug in the upper number (here $x = 1$) and plug in the lower number (here $x = {}^-2$), giving two expressions.
3. Subtract the second expression from the first to achieve the integral value.

Examples:

1. $\int_{-2}^{1} 2x\,dx = x^2 \Big]_{-2}^{1}$ Take the antiderivative of 2.

    $\int_{-2}^{1} 2x\,dx = 1^2 - ({}^-2)^2$ Substitute in $x = 1$ and $x = {}^-2$ and subtract the results.

    $\int_{-2}^{1} 2x\,dx = 1 - 4 = {}^-3$ The integral has the value $^-3$.

2. $\int_{0}^{\frac{\pi}{2}} \cos x\,dx = \sin x \Big]_{0}^{\frac{\pi}{2}}$ The antiderivative of $\cos x$ is $\sin x$.

    $\int_{0}^{\frac{\pi}{2}} \cos x\,dx = \sin \frac{\pi}{2} - \sin 0$ Substitute in $x = \frac{\pi}{2}$ and $x = 0$. Subtract the results.

    $\int_{0}^{\frac{\pi}{2}} \cos x\,dx = 1 - 0 = 1$ The integral has the value 1.

TEACHER CERTIFICATION EXAM

SKILL 55.2 Evaluate integrals by use of basic integration formulas.

A list of integration formulas follows. In each case the letter u is used to represent either a single variable or an expression. Note that also in each case du is required. du is the derivative of whatever u stands for. If u is sin x then du is cos$x$, which is the derivative of sin$x$. If the derivative of u is not entirely present, remember you can put in constants as long as you also insert the reciprocal of any such constants. n is a natural number.

$$\int u^n du = \frac{1}{n+1} u^{n+1} + c \quad \text{if } n \neq {}^-1$$

$$\int \frac{1}{u} du = \ln|u| + c$$

$$\int e^u du = e^u + c$$

$$\int a^u du = \frac{1}{\ln a} a^u + c$$

$$\int \sin u\, du = {}^-\cos u + c$$

$$\int \cos u\, du = \sin u + c$$

$$\int \sec^2 u\, du = \tan u + c$$

$$\int \csc^2 u\, du = {}^-\cot u + c$$

Example:

1. $\int \frac{6}{x} dx = 6 \int \frac{1}{x} dx$ You can pull any constants outside the integral.

   $\int \frac{6}{x} dx = 6\ln|x| + c$

MATHEMATICS HIGH SCHOOL   289

TEACHER CERTIFICATION EXAM

SKILL 55.3 Evaluate an integral by use of substitution.

Sometimes an integral does not always look exactly like the forms from section 55.2. But with a simple substitution (sometimes called a *u* substitution), the integral can be made to look like one of the general forms.

You might need to experiment with different *u* substitutions before you find the one that works. Follow these steps.

1. If the object of the integral is a sum or difference, first split the integral up.
2. For each integral, see if it fits one of the general forms from section 55.2.
3. If the integral does not fit one of the forms, substitute the letter *u* in place of one of the expressions in the integral.
4. Off to the side, take the derivative of *u*, and see if that derivative exists inside the original integral. If it does, replace that derivative expression with *du*. If it does not, try another *u* substitution.
5. Now the integral should match one of the general forms, including both the *u* and the *du*.
6. Take the integral using the general forms, and substitute for the value of *u*.

Examples:

1. $\int (\sin x^2 \cdot 2x + \cos x^2 \cdot 2x) dx$ — Split the integral up.

    $\int \sin(x^2) \cdot 2x\,dx + \int \cos(x^2) 2x\,dx$

    $u = x^2, \quad du = 2x\,dx$ — If you let $u = x^2$, the derivative of *u*, *du*, is $2x\,dx$.

    $\int \sin u\,du + \int \cos u\,du$ — Make the *u* and *du* substitutions.

    $^-\cos u + {}^-\sin u + c$ — Use the formula for integrating sin*u*.

    $^-\cos(x^2) - \sin(x^2) + c$ — Substitute back in for *u*.

2. $\int e^{\sin x} \cos x\,dx$ — Try letting $u = \cos x$. The

    $u = \cos x, \quad du = {}^- \sin x\,dx$ — derivative of *u* would be $^-\sin x\,dx$, which is not present.

    $u = \sin x, \quad du = \cos x\,dx$ — Try another substitution: $u = \sin x$, $du = \cos x\,dx$. $du = \cos x\,dx$ is present.

    $\int e^u\,du$ — e^u is one of the general forms.

    e^u — The integral of e^u is e^u.

    $e^{\sin x}$ — Substitute back in for *u*.

MATHEMATICS HIGH SCHOOL

TEACHER CERTIFICATION EXAM

SKILL 55.4 Evaluate an integral by use of the method of integration by parts.

Integration by parts should only be used if none of the other integration methods works. Integration by parts requires two substitutions (both *u* and *dv*).

1. Let *dv* be the part of the integral that you think can be integrated by itself.
2. Let *u* be the part of the integral which is left after the *dv* substitution is made.
3. Integrate the *dv* expression to arrive at just simply *v*.
4. Differentiate the *u* expression to arrive at *du*. If *u* is just *x*, then *du* is dx.
5. Rewrite the integral using $\int u\,dv = uv - \int v\,du$.
6. All that is left is to integrate $\int v\,du$.
7. If you cannot integrate *v du,* try a different set of substitutions and start the process over.

Examples:

1. $\int xe^{3x}\,dx$ Make *dv* and *u* substitutions.

   $dv = e^{3x}\,dx \quad u = x$ Integrate the *dv* term to arrive at *v*.

   $v = \frac{1}{3}e^{3x} \quad du = dx$ Differentiate the *u* term to arrive at *du*.

   $\int xe^{3x}\,dx = x\left(\frac{1}{3}e^{3x}\right) - \int \frac{1}{3}e^{3x}\,dx$ Rewrite the integral using the above formula.

   $\int xe^{3x}\,dx = \frac{1}{3}xe^{3x} - \frac{1}{3} \cdot \frac{1}{3}\int e^{3x}\,3\,dx$ Before taking the integral of $\frac{1}{3}e^{3x}\,dx$, you must put in a 3 and another 1/3.

   $\int xe^{3x}\,dx = \frac{1}{3}xe^{3x} - \frac{1}{9}e^{3x} + c$ Integrate to arrive at the solution.

2. $\int \ln 4x\,dx$ Note that no other integration method will work.

   $dv = dx \quad u = \ln 4x$ Make the *dv* and *u* substitutions.

   $v = x \quad du = \frac{1}{4x} \cdot 4 = \frac{1}{x}\,dx$ Integrate *dx* to get *x*.

        Differentiate ln4x to get $(1/x)\,dx$.

   $\int \ln 4x\,dx = \ln 4x \cdot x - \int x \cdot \frac{1}{x}\,dx$ Rewrite the formula above.

   $\int \ln 4x\,dx = \ln 4x \cdot x - \int dx$ Simplify the integral.

   $\int \ln 4\,dx = \ln 4x \cdot x - x + c$ Integrate *dx* to get the value $x + c$

MATHEMATICS HIGH SCHOOL

COMPETENCY 56.0   **USE OF INTEGRAL CALCULUS TO FIND THE AREA BETWEEN CURVES AND THE VOLUME OF A SOLID OF REVOLUTION.**

SKILL 56.1   Find the area under a curve by using integration.

Taking the integral of a function and evaluating it from one x value to another provides the total area under the curve (i.e. between the curve and the x axis). Remember, though, that regions above the x axis have "positive" area and regions below the x axis have "negative" area. You must account for these positive and negative values when finding the area under curves. Follow these steps.

1. Determine the x values that will serve as the left and right boundaries of the region.
2. Find all x values between the boundaries that are either solutions to the function or are values which are not in the domain of the function. These numbers are the interval numbers.
3. Integrate the function.
4. Evaluate the integral once for each of the intervals using the boundary numbers.
5. If any of the intervals evaluates to a negative number, make it positive (the negative simply tells you that the region is below the x axis).
6. Add the value of each integral to arrive at the area under the curve.

Example:
Find the area under the following function on the given intervals.

$f(x) = \sin x$; $(0, 2\pi)$

$\sin x = 0$ Find any roots to $f(x)$ on $(0, 2\pi)$.

$x = \pi$

$(0, \pi)$ $(\pi, 2\pi)$ Determine the intervals using the boundary numbers and the roots.

$\int \sin x\, dx = {}^- \cos x$ Integrate $f(x)$. We can ignore the constant c because we have numbers to use to evaluate the integral.

${}^- \cos x \Big]_{x=0}^{x=\pi} = {}^-\cos \pi - ({}^-\cos 0)$

${}^- \cos x \Big]_{x=0}^{x=\pi} = {}^-(-1) + (1) = 2$

${}^- \cos x \Big]_{x=0}^{x=\pi} = {}^-\cos 2\pi - ({}^-\cos 0)$

${}^- \cos x \Big]_{x=0}^{x=\pi} = {}^-1 + ({}^-1) = {}^-2$ The ${}^-2$ means that for $(\pi, 2\pi)$, the region is below the x axis, but the area is still 2.

Area $= 2 + 2 = 4$ Add the 2 integrals together to get the area.

SKILL 56.2 Find the area between two curves by using integration.

Finding the area between two curves is much the same as finding the area under one curve. But instead of finding the roots of the functions, you need to find the *x* values which produce the same number from both functions (set the functions equal and solve). Use these numbers and the given boundaries to write the intervals. On each interval you must pick sample values to determine which function is "on top" of the other. Find the integral of each function. For each interval, subtract the "bottom" integral from the "top" integral. Use the interval numbers to evaluate each of these differences. Add the evaluated integrals to get the total area between the curves.

Example:

Find the area of the regions bounded by the two functions on the indicated intervals.

$f(x) = x + 2$ and $g(x) = x^2$ $[^-2, 3]$ Set the functions equal and solve.

$x + 2 = x^2$
$0 = (x-2)(x+1)$
$x = 2$ or $x = {}^-1$

$(^-2, ^-1)$ $(^-1, 2)$ $(2, 3)$ Use the solutions and the boundary numbers to write the intervals.

$f(^-3/2) = \left(\dfrac{^-3}{2}\right) + 2 = \dfrac{1}{2}$ Pick sample values on the integral and evaluate each function as that number.

$g(^-3/2) = \left(\dfrac{^-3}{2}\right)^2 = \dfrac{9}{4}$ $g(x)$ is "on top" on $[^-2, ^-1]$.

$f(0) = 2$ $f(x)$ is "on top" on $[^-1, 2]$.

$g(0) = 0$

$$f(5/2) = \frac{5}{2} + 2 = \frac{9}{2}$$ g(x) is "on top" on [2,3].

$$g(5/2) = \left(\frac{5}{2}\right)^2 = \frac{24}{4}$$

$$\int f(x)dx = \int (x+2)dx$$
$$\int f(x)dx = \int x\,dx + 2\int dx$$
$$\int f(x)dx = \frac{1}{1+1}x^{1+1} + 2x$$
$$\int f(x)dx = \frac{1}{2}x^2 + 2x$$
$$\int g(x)dx = \int x^2 dx$$
$$\int g(x)dx = \frac{1}{2+1}x^{2+1} = \frac{1}{3}x^3$$

Area 1 $= \int g(x)dx - \int f(x)dx$ g(x) is "on top" on $[^-2,^-1]$.

Area 1 $= \frac{1}{3}x^3 - \left(\frac{1}{2}x^2 + 2x\right)\Big]_{-2}^{-1}$

Area 1 $= \left[\frac{1}{3}(^-1)^3 - \left(\frac{1}{2}(^-1)^2 + 2(^-1)\right)\right] - \left[\frac{1}{3}(^-2)^3 - \left(\frac{1}{2}(^-2)^2 + 2(^-2)\right)\right]$

Area 1 $= \left[\frac{^-1}{3} - \left(\frac{^-3}{2}\right)\right] - \left[\frac{^-8}{3} - (^-2)\right]$

Area 1 $= \left(\frac{7}{6}\right) - \left(\frac{^-2}{3}\right) = \frac{11}{6}$

Area 2 $= \int f(x)dx - \int g(x)dx$ f(x) is "on top" on $[^-1,2]$.

Area 2 $= \frac{1}{2}x^2 + 2x - \frac{1}{3}x^3\Big]_{-1}^{2}$

Area 2 $= \left(\frac{1}{2}(2)^2 + 2(2) - \frac{1}{3}(2)^3\right) - \left(\frac{1}{2}(^-1)^2 + 2(^-1) - \frac{1}{3}(^-1)^3\right)$

Area 2 $= \left(\frac{10}{3}\right) - \left(\frac{1}{2} - 2 + \frac{1}{3}\right)$

Area 2 $= \frac{27}{6}$

Area 3 = $\int g(x)dx - \int f(x)dx$ $\quad\quad$ g(x) is "on top" on [2,3].

Area 3 = $\frac{1}{3}x^3 - \left(\frac{1}{2}x^2 + 2x\right)\Big]_2^3$

Area 3 = $\left[\frac{1}{3}(3)^3 - \left(\frac{1}{2}(3^2) + 2(3)\right)\right] - \left[\frac{1}{3}(2)^3 - \left(\frac{1}{2}(2)^2 + 2(2)\right)\right]$

Area 3 = $\left(\frac{-3}{2}\right) - \left(\frac{-10}{3}\right) = \frac{11}{6}$

Total area = $\frac{11}{6} + \frac{27}{6} + \frac{11}{6} = \frac{49}{6} = 8\frac{1}{6}$

SKILL 56.3 Find the volume of a solid of revolution.

If you take the area bounded by a curve or curves and revolve it about a line, the result is a solid of revolution. To find the volume of such a solid, the Washer Method works in most instances. Imagine slicing through the solid perpendicular to the line of revolution. The "slice" should resemble a washer. Use an integral and the formula for the volume of disk.

$$Volume_{disk} = \pi \cdot radius^2 \cdot thickness$$

Depending on the situation, the radius is the distance from the line of revolution to the curve; or if there are two curves involved, the radius is the difference between the two functions. The thickness is dx if the line of revolution is parallel to the x axis and dy if the line of revolution is parallel to the y axis. Finally, integrate the volume expression using the boundary numbers from the interval.

Example:

Find the value of the solid of revolution found by revolving $f(x) = 9 - x^2$ about the x axis on the interval $[0, 4]$.

$radius = 9 - x^2$
$thickness = dx$

$Volume = \int_0^4 \pi(9-x^2)^2 dx$ Use the formula for volume of a disk.

$Volume = \pi \int_0^4 \left(81 - 18x^2 + x^4\right) dx$

$Volume = \pi \left(81x - \dfrac{18}{2+1}x^3 + \dfrac{1}{4+1}x^5 \right) \Big]_0^4$ Take the integral.

$Volume = \pi \left(81x - 6x^3 + \dfrac{1}{5}x^5 \right) \Big]_0^4$ Evaluate the integral first

$Volume = \pi \left[\left(324 - 384 + \dfrac{1024}{5} \right) - (0 - 0 + 0) \right]$ $x = 4$ then at $x = 0$

$Volume = \pi \left(144 \dfrac{4}{5} \right) = 144 \dfrac{4}{5} \pi$ and subtract.

CURRICULUM AND INSTRUCTION

The National Council of Teachers of Mathematics standards emphasize the teacher's obligation to make mathematics relevant to the students and applicable to the real world. The mathematics need in our technological society is different from that need in the past; we need thinking skills rather than computational. Mathematics needs to connect to other subjects, as well as other areas of math.

ERROR ANALYSIS

A simple method for analyzing student errors is to ask how the answer was obtained. The teacher can then determine if a common error pattern has resulted in the wrong answer. There is a value to having the students explain how the arrived at the correct as well as the incorrect answers.

Many errors are due to simple **carelessness**. Students need to be encouraged to work slowly and carefully. They should check their calculations by redoing the problem on another paper, not merely looking at the work. Addition and subtraction problems need to be written neatly so the numbers line up. Students need to be careful regrouping in subtraction. Students must write clearly and legibly, including erasing fully. Use estimation to ensure that answers make sense.

Many students' computational skills exceed their **reading** level. Although they can understand basic operations, they fail to grasp the concept or completely understand the question. Students must read directions slowly.

Fractions are often a source of many errors. Students need to be reminded to use common denominators when adding and subtracting and to always express answers in simplest terms. Again, it is helpful to check by estimating.

The most common error that is made when working with **decimals** is failure to line up the decimal points when adding or subtracting or not moving the decimal point when multiplying or dividing. Students also need to be reminded to add zeroes when necessary. Reading aloud may also be beneficial. Estimation, as always, is especially important.

Students need to know that it is okay to make mistakes. The teacher must keep a positive attitude, so they do not feel defeated or frustrated.

REPRESENTATIONS OF CONCEPTS

Mathematical operations can be shown using manipulatives or drawings. Multiplication can be shown using arrays.

3×4

Addition and subtractions can be demonstrated with symbols.

ψ ψ ψ ξ ξ ξ ξ
3 + 4 = 7
7 - 3 = 4

Fractions can be clarifies using pattern blocks, fraction bars, or paper folding.

CONCEPT DEVELOPMENT

Manipulatives can foster learning for all students. Mathematics needs to be derived from something that is real to the learner. If he can "touch" it, he will understand and remember. Students can use fingers, ice cream sticks, tiles and paper folding, as well as those commercially available manipulatives to visualize operations and concepts. The teacher needs to solidify the concrete examples into abstract mathematics.

TEACHER CERTIFICATION EXAM

PROBLEM SOLVING

Problem solving strategies are simply plans of attack. Student often panic when confronted with word problems. If they have a "list" of ideas, ways to attempt a solution, they will be able to approach the problems more calmly and confidently. Some methods include, but are not limited to, draw a diagram, work backwards, guess and check, and solve a simpler problem.

It is helpful to have students work in groups. Mathematics does not have to be solitary activity. Cooperative learning fosters enthusiasm. Creating their own problems is another useful tool. Also, encourage students to find more than one way to solve a problem. Thinking about problem solving after the solution has been discovered encourages understanding and creativity. The more they practice problems, the more comfortable and positive students will feel.

MATHEMATICAL LANGUAGE

Students need to use the proper mathematical terms and expressions. When reading decimals, they need to read 0.4 as "four tenths" to promote better understanding of the concepts. They should do their work in a neat and organized manner. Students need to be encouraged to verbalize their strategies, both in computation and word problems. Additionally, writing original word problems fosters understanding of math language. Another idea is requiring students to develop their own glossary of mathematical glossary. Knowing the answers and being able to communicate them are equally important.

MATHEMATICS HIGH SCHOOL

MANIPULATIVES

Example:
Using tiles to demonstrate both geometric ideas and number theory.

Give each group of students 12 tiles and instruct them to build rectangles. Students draw their rectangles on paper.

12 × 1

1 × 12

3 × 4

4 × 3

6 × 2

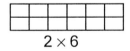

2 × 6

Encourage students to describe their reactions. Extend to 16 tiles. Ask students to form additional problems.

CALCULATORS

Calculators are an important tool. They should be encouraged in the classroom and at home. They do not replace basic knowledge but they can relieve the tedium of mathematical computations, allowing students to explore more challenging mathematical directions. Students will be able to use calculators more intelligently if they are taught how. Students need to always check their work by estimating. The goal of mathematics is to prepare the child to survive in the real world. Technology is a reality in today's society.

CHILD DEVELOPMENT

Means of instruction need to be varied to reach children of different skill levels and learning styles. In addition to directed instruction, students should work cooperatively, explore hands-on activities and do projects.

COMPUTERS

Computers can not replace teachers. However, they can be used to enhance the curriculum. They may be used cautiously to help students practice basic skills. Many excellent programs exist to encourage higher-order thinking skills, creativity and problem solving. Learning to use technology appropriately is an important preparation for adulthood. Computers can also show the connections between mathematics and the real world.

QUESTIONING TECHNIQUES

As the teacher's role in the classroom changes from lecturer to facilitator, the questions need to further stimulate students in various ways.
- Helping students work together

What do you think about what John said?
Do you agree? Disagree?
Can anyone explain that differently?

- Helping students determine for themselves if an answer is correct

What do you think that is true?
How did you get that answer?
Do you think that is reasonable? Why?

- Helping students learn to reason mathematically

Will that method always work?
Can you think of a case where it is not true?
How can you prove that?
Is that answer true in all cases?

- Helping student brainstorm and problem solve

Is there a pattern?
What else can you do?
Can you predict the answer?
What if...?

- Helping students connect mathematical ideas

What did we learn before that is like this?
Can you give an example?
What math did you see on television last night? in the newspaper?

TEACHER CERTIFICATION EXAM

ANSWER KEY TO PRACTICE PROBLEMS

Skill 8.1

    Question #2 a, b, c, f are functions
Question #3 Domain = $^-\infty, \infty$ Range = $^-5, \infty$

Skill 8.2

    Question #1 Domain = x Range = $y \geq ^-6$
Question #2 Domain = 1,4,7,6   Range = -2
Question #3 Domain = $x \neq 2, ^-2$
Question #4 Domain = $^-\infty, \infty$ Range = -4, 4
                Domain = $^-\infty, \infty$ Range = 2, ∞
Question #5 Domain = $^-\infty, \infty$ Range = $\varnothing$
Question #6 (3,9), (-4,16), (6,3), (1,9), (1,3)

Skill 8.3

    Question #1

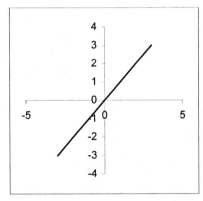

    Question #2

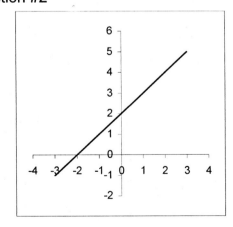

MATHEMATICS HIGH SCHOOL

Question #3

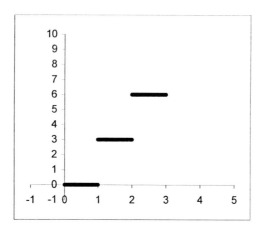

Questions #4

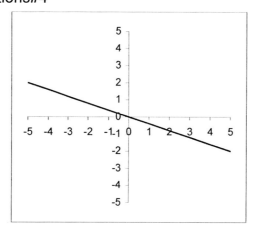

Skill 8.4

Question #1

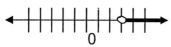

Question #2

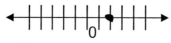

Question #3

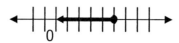

Question #4

Skill 8.5

Question #1

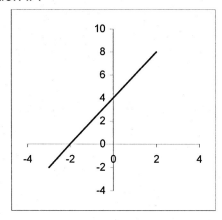

Question #2

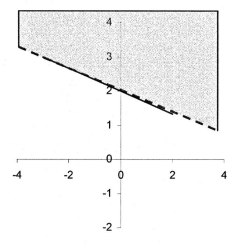

Question #3

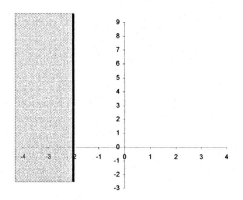

TEACHER CERTIFICATION EXAM

Skill 8.6

Question #1   x-intercept = -14   y-intercept = -10   slope = $-\dfrac{5}{7}$

Question #2   x-intercept = 14   y-intercept = -7   slope = $\dfrac{1}{2}$

Question #3   x-intercept = 3   y-intercept = none

Question #4   x-intercept = $\dfrac{15}{2}$ y-intercept = 3   slope = $-\dfrac{2}{5}$

Skill 8.8

Question #1 $y = \dfrac{3}{4}x + \dfrac{17}{4}$

Question #2 $x = 11$

Question #3 $y = \dfrac{3}{5}x + \dfrac{42}{5}$

Question #4 $y = 5$

Skill 9.1

Question #1 $\dfrac{^-12x^5 z^5}{y}$

Question #2 $25a^4 + 15a^2 - 7b^3$

Question #3 $100x^{20}y^8 + 8x^9 y^6$

Skill 10.1

Question #5 $\dfrac{8x+36}{(x-3)(x+7)}$

Question #6 $\dfrac{25a^2 + 12b^2}{20a^4 b^5}$

Question #7 $\dfrac{2x^2 + 5x - 21}{(x-5)(x+5)(x+3)}$

TEACHER CERTIFICATION EXAM

Skill 10.2

Question #1 It takes Curly 15 minutes to paint the elephant alone
Question #2 The original number is 5
Question #3 The car was travelling at 68mph and the truck was at 62mph

Skill 10.3

Question #1 $C = \dfrac{5}{9}F - \dfrac{160}{9}$

Question #2 $b = \dfrac{2A - 2h^2}{h}$

Question #3 $n = \dfrac{360 + S}{180}$

Skill 11.1

Question #1 $(6x - 5y)(36x^2 + 30xy + 25y^2)$

Question #2 $4(a - 2b)(a^2 + 2ab + 4b^2)$

Question #3 $5x^2(2x^9 + 3y)(4x^{18} - 6x^9 y + 9y^2)$

Skill 11.2

Question #1 $(2x - 5y)(2x + 5y)$

Question #2 $2(3b - 4)(b + 1)$

Question #3 The answer is D (2x+1)

Skill 12.1

Question #1 $9xz^5$

Question #2 $\dfrac{3x + 2y}{x^2 + 5xy + 25y^2}$

Question #3 $\dfrac{x^2 + 8x + 15}{(x+2)(x+3)(x-7)}$

TEACHER CERTIFICATION EXAM

Skill 12.2

Question #1 $\dfrac{14x+28}{(x+6)(x+1)(x-1)}$

Question #2 $\dfrac{x^3-5x^2+10x-12}{x^2+3x-10}$

Skill 12.3

Question #1 $x=7, x=^-5$

Question #2 $x=\dfrac{13}{8}$

Skill 12.4

Question #1 $6a^4\sqrt{2a}$

Question #2 $7i\sqrt{2}$

Question #3 $2x^2 i$

Question #4 $6x^4 y^3 \sqrt[4]{3x^2 y^3}$

Skill 12.5

Question #1 $5\sqrt{6}$

Question #2 $(6+\sqrt{3})(^-5+\sqrt{5})$

Question #3 $\dfrac{12+3\sqrt{6}+4\sqrt{3}+3\sqrt{2}}{5}$

Skill 12.6

Question #1 $17\sqrt{6}$

Question #2 $84\sqrt{2}x^4 y^8$

Question #3 $\dfrac{5a^4\sqrt{35b}}{8b}$

Question #4 $^-15-5\sqrt{3}-10\sqrt{5}$

Skill 12.7

Question #1 $x=11$

Question #2 $x=7, x=23$

Question #3 $x=17$

MATHEMATICS HIGH SCHOOL

TEACHER CERTIFICATION EXAM

Skill 13.2

Question #1   The sides are 8, 15, and 17

Question #2   The numbers are 2 and $\dfrac{1}{2}$

Skill 13.3

Question #1 $x = \dfrac{7 \pm \sqrt{241}}{12}$

Question #2 $x = \dfrac{1}{2}$ or $\dfrac{-2}{5}$

Question #3 $x = \dfrac{8}{5}$

Skill 13.4

Question #1 $x^2 - 10x + 25$

Question #2 $25x^2 - 10x - 48$

Question #3 $x^2 - 9x - 36$

Skill 20.3

Question #1 $S_5 = 75$

Question #2 $S_n = 28$

Question #3 $S_n = 2186$

Question #4 $S_n = -1.99$

Skill 22.2

Question #1 $35x^3 y^4$

Question #2 $78732 x^7 y^2$

Skill 30.0

Question #1   The Red Sox won the World Series.
Question #2   Angle B is not between 0 and 90 degrees.
Question #3   Annie will do well in college.
Question #4   You are witty and charming.

Skill 35.3

Question #1 $\begin{pmatrix} -4 & 0 & -2 \\ 2 & 4 & -8 \end{pmatrix}$

Question #2 $\begin{pmatrix} 18 \\ 34 \\ 32 \end{pmatrix}$

Question #3 $\begin{pmatrix} -12 & 16 \\ -4 & -2 \\ 0 & 6 \end{pmatrix}$

Skill 35.4

Question #1 $\begin{pmatrix} -7 & 0 \\ -12 & -3 \\ 6 & 5 \end{pmatrix}$

Question #2 $\begin{pmatrix} 3 & \frac{47}{4} \\ 3 & 7 \end{pmatrix}$

Question #3 a = -5   b = -1   c = 0   d = -4   e = 0   f = -10

Skill 35.5

Question #1 $\begin{pmatrix} -15 & 25 \\ -1 & -13 \end{pmatrix}$

Question #2 $\begin{pmatrix} 5 & -5 & -10 \\ 5 & 8 & 0 \\ 1 & 5 & 7 \\ -9 & 13 & 22 \end{pmatrix}$

TEACHER CERTIFICATION EXAM

Skill 35.6

Question #1 $\begin{pmatrix} x \\ y \end{pmatrix} = \begin{pmatrix} 3 \\ 1 \end{pmatrix}$

Question #2 $\begin{pmatrix} x \\ y \\ z \end{pmatrix} = \begin{pmatrix} 4 \\ 4 \\ 1 \end{pmatrix}$

Skill 37.1

Question #1
$$\cot\theta = \frac{x}{y} \qquad \cos\theta = \frac{x}{r} \qquad \sin\theta = \frac{y}{r}$$
$$\frac{x}{y} = \frac{x}{r} \times \frac{r}{y} = \frac{x}{y} = \cot\theta$$

Question #2
$$1 + \cot^2\theta = \csc^2\theta$$
$$\frac{y^2}{y^2} + \frac{x^2}{y^2} = \frac{r^2}{y^2} = \csc^2\theta$$

Skill 45.1

Question #1

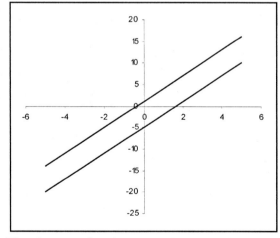

The locus of points are those points on and between the two lines.

MATHEMATICS HIGH SCHOOL

TEACHER CERTIFICATION EXAM

Skill 48.2

    Question #1 $x = 2.2$
    Question #2
    Question #3

Skill 49.1

    Question #1   Approx. 49.358
    Question #2   error-There is no answer because zero is in the denominator

Skill 49.2

    Question #1 ∞
    Question #2   -1

Skill 53.2

    Question #1 $g(x) = -20\sin x + c$
    Question #2 $g(x) = \pi \sec x + c$

Skill 54.1

    Question #1 $t(0) = -24$ m/sec
    Question #2 $t(4) = 24$ m/sec

TEACHER CERTIFICATION EXAM

1. Change $.\overline{63}$ into a fraction in simplest form.

   A) 63/100
   B) 7/11
   C) 6 3/10
   D) 2/3

2. Which of the following sets is closed under division?

   I) {½, 1, 2, 4}
   II) {-1, 1}
   III) {-1, 0, 1}

   A) I only
   B) II only
   C) III only
   D) I and II

3. Which of the following illustrates an inverse property?

   A) a + b = a - b
   B) a + b = b + a
   C) a + 0 = a
   D) a + (-a) = 0

4. $f(x) = 3x - 2$; $f^{-1}(x) =$

   A) $3x + 2$
   B) $x/6$
   C) $2x - 3$
   D) $(x+2)/3$

5. What would be the total cost of a suit for $295.99 and a pair of shoes for $69.95 including 6.5% sales tax?

   A) $389.73
   B) $398.37
   C) $237.86
   D) $315.23

6. A student had 60 days to appeal the results of an exam. If the results were received on March 23, what was the last day that the student could appeal?

   A) May 21
   B) May 22
   C) May 23
   D) May 24

7. Which of the following is always composite if x is odd, y is even, and both x and y are greater than or equal to 2?

   A) $x + y$
   B) $3x + 2y$
   C) $5xy$
   D) $5x + 3y$

8. Which of the following is incorrect?

   A) $(x^2 y^3)^2 = x^4 y^6$
   B) $m^2(2n)^3 = 8m^2 n^3$
   C) $(m^3 n^4)/(m^2 n^2) = mn^2$
   D) $(x + y^2)^2 = x^2 + y^4$

9. Express .0000456 in scientific notation.

   A) 4.56×10^{-4}
   B) 45.6×10^{-6}
   C) 4.56×10^{-6}
   D) 4.56×10^{-5}

MATHEMATICS HIGH SCHOOL

10. Compute the area of the shaded region, given a radius of 5 meters. 0 is the center.

    A) 7.13 cm²
    B) 7.13 m²
    C) 78.5 m²
    D) 19.63 m²

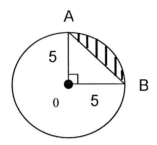

11. If the area of the base of a cone is tripled, the volume will be

    A) the same as the original
    B) 9 times the original
    C) 3 times the original
    D) 3 π times the original

12. Find the area of the figure pictured below.

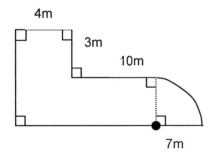

    A) 136.47 m²
    B) 148.48 m²
    C) 293.86 m²
    D) 178.47 m²

13. The mass of a Chips Ahoy cookie would be approximately

    A) 1 kilogram
    B) 1 gram
    C) 15 grams
    D) 15 milligrams

14. Compute the median for the following data set:

    {12, 19, 13, 16, 17, 14}

    A) 14.5
    B) 15.17
    C) 15
    D) 16

15. Half the students in a class scored 80% on an exam, most of the rest scored 85% except for one student who scored 10%. Which would be the best measure of central tendency for the test scores?

    A) mean
    B) median
    C) mode
    D) either the median or the mode because they are equal

16. What conclusion can be drawn from the graph below?

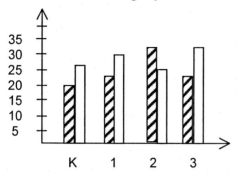

MLK Elementary Student Enrollment Girls Boys

A) The number of students in first grade exceeds the number in second grade.
B) There are more boys than girls in the entire school.
C) There are more girls than boys in the first grade.
D) Third grade has the largest number of students.

17) State the domain of the function $f(x) = \dfrac{3x-6}{x^2 - 25}$

A) $x \neq 2$
B) $x \neq 5, -5$
C) $x \neq 2, -2$
D) $x \neq 5$

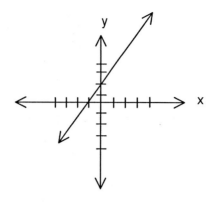

18. What is the equation of the above graph?

A) $2x + y = 2$
B) $2x - y = -2$
C) $2x - y = 2$
D) $2x + y = -2$

19. Solve for v_0: $d = at(v_t - v_0)$

A) $v_0 = atd - v_t$
B) $v_0 = d - atv_t$
C) $v_0 = atv_t - d$
D) $v_0 = (atv_t - d)/at$

20. Which of the following is a factor of $6 + 48m^3$

A) (1 + 2m)
B) (1 - 8m)
C) (1 + m - 2m)
D) (1 - m + 2m)

TEACHER CERTIFICATION EXAM

21. Which graph represents the equation of $y = x^2 + 3x$?

A)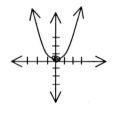
B)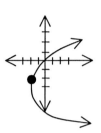

C)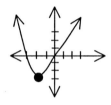
D)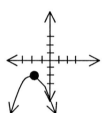

22. The volume of water flowing through a pipe varies directly with the square of the radius of the pipe. If the water flows at a rate of 80 liters per minute through a pipe with a radius of 4 cm, at what rate would water flow through a pipe with a radius of 3 cm?

A) 45 liters per minute
B) 6.67 liters per minute
C) 60 liters per minute
D) 4.5 liters per minute

23) Solve the system of equations for x, y and z.

$3x + 2y - z = 0$
$2x + 5y = 8z$
$x + 3y + 2z = 7$

A) $(-1, 2, 1)$
B) $(1, 2, -1)$
C) $(-3, 4, -1)$
D) $(0, 1, 2)$

24. Solve for x: $18 = 4 + |2x|$

A) $\{-11, 7\}$
B) $\{-7, 0, 7\}$
C) $\{-7, 7\}$
D) $\{-11, 11\}$

25. Which graph represents the solution set for $x^2 - 5x > -6$?

A) ←—○———○—→
       -2 0 2

B) ←○———————○→
     -3   0   3

C) ←—○———○—→
       -2 0 2

D) ←←————○○——→
     -3   0 2 3

26. Find the zeroes of
$f(x) = x^3 + x^2 - 14x - 24$

A) 4, 3, 2
B) 3, -8
C) 7, -2, -1
D) 4, -3, -2

27. Evaluate $3^{1/2}(9^{1/3})$

A) $27^{5/6}$
B) $9^{7/12}$
C) $3^{5/6}$
D) $3^{6/7}$

28. Simplify: $\sqrt{27} + \sqrt{75}$

A) $8\sqrt{3}$
B) 34
C) $34\sqrt{3}$
D) $15\sqrt{3}$

29. Simplify: $\dfrac{10}{1+3i}$

A) $-1.25(1-3i)$
B) $1.25(1+3i)$
C) $1+3i$
D) $1-3i$

30. Find the sum of the first one hundred terms in the progression.
(-6, -2, 2 . . .)

A) 19,200
B) 19,400
C) -604
D) 604

31. How many ways are there to choose a potato and two green vegetables from a choice of three potatoes and seven green vegetables?

A) 126
B) 63
C) 21
D) 252

32. What would be the seventh term of the expanded binomial $(2a+b)^8$?

A) $2ab^7$
B) $41a^4b^4$
C) $112a^2b^6$
D) $16ab^7$

33. Which term most accurately describes two coplanar lines without any common points?

A) perpendicular
B) parallel
C) intersecting
D) skew

34. Determine the number of subsets of set K.
K = {4, 5, 6, 7}

A) 15
B) 16
C) 17
D) 18

35. What is the degree measure of an interior angle of a regular 10 sided polygon?

A) 18°
B) 36°
C) 144°
D) 54°

36. If a ship sails due south 6 miles, then due west 8 miles, how far was it from the starting point?

A) 100 miles
B) 10 miles
C) 14 miles
D) 48 miles

37. What is the measure of minor arc AD, given measure of arc PS is 40° and $m\angle K = 10°$?

   A) 50°
   B) 20°
   C) 30°
   D) 25°

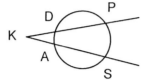

38. Choose the diagram which illustrates the construction of a perpendicular to the line at a given point on the line.

   A)

   B)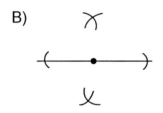

   C)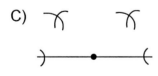

   D)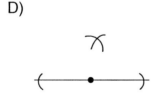

39. When you begin by assuming the conclusion of a theorem is false, then show that through a sequence of logically correct steps you contradict an accepted fact, this is known as

   A) inductive reasoning
   B) direct proof
   C) indirect proof
   D) exhaustive proof

40. Which theorem can be used to prove $\triangle BAK \cong \triangle MKA$?

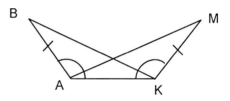

   A) SSS
   B) ASA
   C) SAS
   D) AAS

41. Given that QO⊥NP and QO=NP, quadrilateral NOPQ can most accurately be described as a

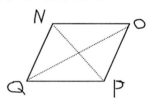

   A) parallelogram
   B) rectangle
   C) square
   D) rhombus

42. Choose the correct statement concerning the median and altitude in a triangle.

   A) The median and altitude of a triangle may be the same segment.
   B) The median and altitude of a triangle are always different segments.
   C) The median and altitude of a right triangle are always the same segment.
   D) The median and altitude of an isosceles triangle are always the same segment.

43. Which mathematician is best known for his work in developing non-Euclidean geometry?

   A) Descartes
   B) Riemann
   C) Pascal
   D) Pythagoras

44. Find the surface area of a box which is 3 feet wide, 5 feet tall, and 4 feet deep.

   A) 47 sq. ft.
   B) 60 sq. ft.
   C) 94 sq. ft
   D) 188 sq. ft.

45. Given a 30 meter x 60 meter garden with a circular fountain having a 5 meter radius, calculate the area of the portion of the garden not occupied by the fountain.

   A) 1721 m²
   B) 1879 m²
   C) 2585 m²
   D) 1015 m²

46. Determine the area of the shaded region of the trapezoid in terms of x and y.

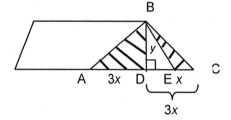

   A) $4xy$
   B) $2xy$
   C) $3x^2 y$
   D) There is not enough information given.

47. Compute the standard deviation for the following set of temperatures.
    (37, 38, 35, 37, 38, 40, 36, 39)

   A) 37.5
   B) 1.5
   C) 0.5
   D) 2.5

MATHEMATICS HIGH SCHOOL

TEACHER CERTIFICATION EXAM

48. Find the value of the determinant of the matrix.

$$\begin{vmatrix} 2 & 1 & -1 \\ 4 & -1 & 4 \\ 0 & -3 & 2 \end{vmatrix}$$

A) 0
B) 23
C) 24
D) 40

49. Which expression is equivalent to $1-\sin^2 x$?

A) $1-\cos^2 x$
B) $1+\cos^2 x$
C) $1/\sec x$
D) $1/\sec^2 x$

50. Determine the measures of angles A and B.

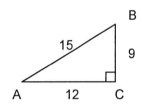

A) A = 30°, B = 60°
B) A = 60°, B = 30°
C) A = 53°, B = 37°
D) A = 37°, B = 53°

51. Given $f(x) = 3x-2$ and $g(x) = x^2$, determine $g(f(x))$.

A) $3x^2 - 2$
B) $9x^2 + 4$
C) $9x^2 - 12x + 4$
D) $3x^3 - 2$

52. Determine the rectangular coordinates of the point with polar coordinates (5, 60°).

A) (0.5, 0.87)
B) (-0.5, 0.87)
C) (2.5, 4.33)
D) (25, 150°)

53. Given a vector with horizontal component 5 and vertical component 6, determine the length of the vector.

A) 61
B) $\sqrt{61}$
C) 30
D) $\sqrt{30}$

54. Compute the distance from (-2,7) to the line x = 5.

A) -9
B) -7
C) 5
D) 7

55. Given $K(-4, y)$ and $M(2,-3)$ with midpoint $L(x,1)$, determine the values of x and y.

A) $x = -1, y = 5$
B) $x = 3, y = 2$
C) $x = 5, y = -1$
D) $x = -1, y = -1$

MATHEMATICS HIGH SCHOOL

TEACHER CERTIFICATION EXAM

56. Find the length of the major axis of $x^2 + 9y^2 = 36$.

    A) 4
    B) 6
    C) 12
    D) 8

57. Which equation represents a circle with a diameter whose endpoints are $(0, 7)$ and $(0, 3)$?

    A) $x^2 + y^2 + 21 = 0$
    B) $x^2 + y^2 - 10y + 21 = 0$
    C) $x^2 + y^2 - 10y + 9 = 0$
    D) $x^2 - y^2 - 10y + 9 = 0$

58. Which equation corresponds to the logarithmic statement: $\log_x k = m$?

    A) $x^m = k$
    B) $k^m = x$
    C) $x^k = m$
    D) $m^x = k$

59. Find the first derivative of the function: $f(x) = x^3 - 6x^2 + 5x + 4$

    A) $3x^3 - 12x^2 + 5x = f'(x)$
    B) $3x^2 - 12x - 5 = f'(x)$
    C) $3x^2 - 12x + 9 = f'(x)$
    D) $3x^2 - 12x + 5 = f'(x)$

60. Differentiate: $y = e^{3x+2}$

    A) $3e^{3x+2} = y'$
    B) $3e^{3x} = y'$
    C) $6e^3 = y'$
    D) $(3x+2)e^{3x+1} = y'$

61. Find the slope of the line tangent to $y = 3x(\cos x)$ at $(\pi/2, \pi/2)$.

    A) $-3\pi/2$
    B) $3\pi/2$
    C) $\pi/2$
    D) $-\pi/2$

62. Find the equation of the line tangent to $y = 3x^2 - 5x$ at $(1, -2)$.

    A) $y = x - 3$
    B) $y = 1$
    C) $y = x + 2$
    D) $y = x$

63. How does the function $y = x^3 + x^2 + 4$ behave from $x = 1$ to $x = 3$?

    A) increasing, then decreasing
    B) increasing
    C) decreasing
    D) neither increasing nor decreasing

MATHEMATICS HIGH SCHOOL

64. Find the absolute maximum obtained by the function $y = 2x^2 + 3x$ on the interval $x = 0$ to $x = 3$.

    A) −3/4
    B) −4/3
    C) 0
    D) 27

65. Find the antiderivative for $4x^3 - 2x + 6 = y$.

    A) $x^4 - x^2 + 6x + C$
    B) $x^4 - 2/3x^3 + 6x + C$
    C) $12x^2 - 2 + C$
    D) $4/3x^4 - x^2 + 6x + C$

66. Find the antiderivative for the function $y = e^{3x}$.

    A) $3x(e^{3x}) + C$
    B) $3(e^{3x}) + C$
    C) $1/3(e^x) + C$
    D) $1/3(e^{3x}) + C$

67. The acceleration of a particle is dv/dt = 6 m/s². Find the velocity at t=10 given an initial velocity of 15 m/s.

    A) 60 m/s
    B) 150 m/s
    C) 75 m/s
    D) 90 m/s

68. If the velocity of a body is given by v = 16 - t², find the distance traveled from t = 0 until the body comes to a complete stop.

    A) 16
    B) 43
    C) 48
    D) 64

69. Evaluate: $\int (x^3 + 4x - 5)dx$

    A) $3x^2 + 4 + C$
    B) $x^4/4 + 2x^2 - 5x + C$
    C) $x^4/3 + 4x - 5x + C$
    D) $x^3 + 4x^2 - 5x + C$

70. Evaluate $\int_0^2 (x^2 + x - 1)dx$

    A) 11/3
    B) 8/3
    C) -8/3
    D) -11/3

71. Find the area under the function $y = x^2 + 4$ from $x = 3$ to $x = 6$.

    A) 75
    B) 21
    C) 96
    D) 57

72. -3 + 7 = -4   6(-10) = - 60
    -5(-15) = 75   -3+-8 = 11
    8-12 = -4   7- -8 = 15

   Which best describes the type of error observed above?

   A) The student is incorrectly multiplying integers.
   B) The student has incorrectly applied rules for adding integers to subtracting integers.
   C) The student has incorrectly applied rules for multiplying integers to adding integers.
   D) The student is incorrectly subtracting integers.

73. **About two weeks after introducing formal proofs, several students in your geometry class are having a difficult time remembering the names of the postulates. They cannot complete the reason column of the proof and as a result are not even attempting the proofs. What would be the best approach to help students understand the nature of geometric proofs?**

   A) Give them more time; proofs require time and experience.
   B) Allow students to write an explanation of the theorem in the reason column instead of the name.
   C) Have the student copy each theorem in a notebook.
   D) Allow the students to have open book tests.

74. **What would be the shortest method of solution for the system of equations below?**

   $3x + 2y = 38$

   $4x + 8 = y$

   A) linear combination
   B) additive inverse
   C) substitution
   D) graphing

75. **Identify the correct sequence of subskills required for solving and graphing inequalities involving absolute value in one variable, such as $|x+1| \leq 6$.**

   A) understanding absolute value, graphing inequalities, solving systems of equations
   B) graphing inequalities on a Cartesian plane, solving systems of equations, simplifying expressions with absolute value
   C) plotting points, graphing equations, graphing inequalities
   D) solving equations with absolute value, solving inequalities, graphing conjunctions and disjunctions

76. What would be the least appropriate use for handheld calculators in the classroom?

   A) practice for standardized tests
   B) integrating algebra and geometry with applications
   C) justifying statements in geometric proofs
   D) applying the law of sines to find dimensions

77. According to Piaget, what stage in a student's development would be appropriate for introducing abstract concepts in geometry?

   A) concrete operational
   B) formal operational
   C) sensori-motor
   D) pre-operational

78. A group of students working with trigonometric identities have concluded that $\cos 2x = 2\cos x$. How could you best lead them to discover their error?

   A) Have the students plug in values on their calculators.
   B) Direct the student to the appropriate chapter in the text.
   C) Derive the correct identity on the board.
   D) Provide each student with a table of trig identities.

79. Which of the following is the best example of the value of personal computers in advanced high school mathematics?

   A) Students can independently drill and practice test questions.
   B) Students can keep an organized list of theorems and postulates on a word processing program.
   C) Students can graph and calculate complex functions to explore their nature and make conjectures.
   D) Students are better prepared for business because of mathematics computer programs in high school.

80. Given the series of examples below, what is $5 \not c 4$?

   $4 \not c 3 = 13 \qquad 7 \not c 2 = 47$
   $3 \not c 1 = 8 \qquad 1 \not c 5 = -4$

   A) 20
   B) 29
   C) 1
   D) 21

TEACHER CERTIFICATION EXAM

ANSWER KEY MATH HIGH SCHOOL SAMPLE TEST

1)	B	31)	A	61)	A
2)	B	32)	C	62)	A
3)	D	33)	B	63)	B
4)	D	34)	B	64)	D
5)	A	35)	C	65)	A
6)	B	36)	B	66)	D
7)	C	37)	B	67)	C
8)	D	38)	D	68)	B
9)	D	39)	C	69)	B
10)	B	40)	C	70)	B
11)	C	41)	C	71)	A
12)	B	42)	A	72)	C
13)	C	43)	B	73)	B
14)	C	44)	C	74)	C
15)	B	45)	A	75)	D
16)	B	46)	B	76)	C
17)	B	47)	B	77)	B
18)	B	48)	C	78)	A
19)	D	49)	D	79)	C
20)	A	50)	D	80)	D
21)	C	51)	C		
22)	A	52)	C		
23)	A	53)	B		
24)	C	54)	D		
25)	D	55)	A		
26)	D	56)	C		
27)	B	57)	B		
28)	A	58)	A		
29)	D	59)	D		
30)	A	60)	A		

MATHEMATICS HIGH SCHOOL

TEACHER CERTIFICATION EXAM

Math High School Sample Test Solutions

1. Let N = .636363.... Then multiplying both sides of the equation by 100 or 10^2 (because there are 2 repeated numbers), we get 100N = 63.636363... Then subtracting the two equations gives 99N = 63 or N = $\frac{63}{99} = \frac{7}{11}$. **Answer is B**

2. I is not closed because $\frac{4}{.5} = 8$ and 8 is not in the set.

   III is not closed because $\frac{1}{0}$ is undefined.

   II is closed because $\frac{-1}{1} = -1, \frac{1}{-1} = -1, \frac{1}{1} = 1, \frac{-1}{-1} = 1$ and all the answers are in the set. **Answer is B**

3. **Answer is D** because a + (-a) = 0 is a statement of the Additive Inverse Property of Algebra.

4. To find the inverse, $f^{-1}(x)$, of the given function, reverse the variables in the given equation, y = 3x – 2, to get x = 3y – 2. Then solve for y as follows: x+2 = 3y, and y = $\frac{x+2}{3}$. **Answer is D.**

5. Before the tax, the total comes to $365.94. Then .065(365.94) = 23.79. With the tax added on, the total bill is 365.94 + 23.79 = $389.73. (Quicker way: 1.065(365.94) = 389.73.) **Answer is A**

6. Recall: 30 days in April and 31 in March. 8 days in March + 30 days in April + 22 days in May brings him to a total of 60 days on May 22. **Answer is B.**

7. A composite number is a number which is not prime. The prime number sequence begins 2,3,5,7,11,13,17,.... To determine which of the expressions is <u>always</u> composite, experiment with different values of x and y, such as x=3 and y=2, or x=5 and y=2. It turns out that 5xy will always be an even number, and therefore, composite, if y=2.
Answer is C.

8. Using FOIL to do the expansion, we get $(x + y^2)^2 = (x + y^2)(x + y^2) = x^2 + 2xy^2 + y^4$. **Answer is D.**

MATHEMATICS HIGH SCHOOL

9. In scientific notation, the decimal point belongs to the right of the 4, the first significant digit. To get from 4.56 x 10^{-5} back to 0.0000456, we would move the decimal point 5 places to the left.
Answer is D.

10. Area of triangle AOB is .5(5)(5) = 12.5 square meters. Since $\frac{90}{360} = .25$, the area of sector AOB (pie-shaped piece) is approximately .25(π)5^2 = 19.63. Subtracting the triangle area from the sector area to get the area of segment AB, we get approximately 19.63-12.5 = 7.13 square meters.
Answer is B.

11. The formula for the volume of a cone is V = $\frac{1}{3}Bh$, where B is the area of the circular base and h is the height. If the area of the base is tripled, the volume becomes V = $\frac{1}{3}(3B)h = Bh$, or three times the original area. **Answer is C.**

12. Divide the figure into 2 rectangles and one quarter circle. The tall rectangle on the left will have dimensions 10 by 4 and area 40. The rectangle in the center will have dimensions 7 by 10 and area 70. The quarter circle will have area .25(π)7^2 = 38.48. The total area is therefore approximately 148.48. **Answer is B.**

13. Since an ordinary cookie would not weigh as much as 1 kilogram, or as little as 1 gram or 15 milligrams, the only reasonable answer is 15 grams. **Answer is C.**

14. Arrange the data in ascending order: 12,13,14,16,17,19. The median is the middle value in a list with an odd number of entries. When there is an even number of entries, the median is the mean of the two center entries. Here the average of 14 and 16 is 15.
Answer is C.

15. In this set of data, the median (see #14) would be the most representative measure of central tendency since the median is independent of extreme values. Because of the 10% outlier, the mean (average) would be disproportionately skewed. In this data set, it is true that the median and the mode (number which occurs most often) are the same, but the median remains the best choice because of its special properties. **Answer is B.**

16. In Kindergarten, first grade, and third grade, there are more boys than girls. The number of extra girls in grade two is more than made up for by the extra boys in all the other grades put together. **Answer is B.**

MATHEMATICS HIGH SCHOOL

TEACHER CERTIFICATION EXAM

17. The values of 5 and –5 must be omitted from the domain of all real numbers because if x took on either of those values, the denominator of the fraction would have a value of 0, and therefore the fraction would be undefined. **Answer is B.**

18. By observation, we see that the graph has a y-intercept of 2 and a slope of 2/1 = 2. Therefore its equation is y = mx + b = 2x + 2. Rearranging the terms gives 2x – y = -2. **Answer is B.**

19. Using the Distributive Property and other properties of equality to isolate v_0 gives d = atv_t – atv_0, atv_0 = atv_t – d, $v_0 = \frac{atv_t - d}{at}$. **Answer is D.**

20. Removing the common factor of 6 and then factoring the sum of two cubes gives 6 + 48m^3 = 6(1 + 8m^3) = 6(1 + 2m)(1^2 – 2m + $(2m)^2$). **Answer is A.**

21. B is not the graph of a function. D is the graph of a parabola where the coefficient of x^2 is negative. A appears to be the graph of y = x^2. To find the x-intercepts of y = x^2 + 3x, set y = 0 and solve for x: 0 = x^2 + 3x = x(x + 3) to get x = 0 or x = -3. Therefore, the graph of the function intersects the x-axis at x=0 and x=-3.
Answer is C.

22. Set up the direct variation: $\frac{V}{r^2} = \frac{V}{r^2}$. Substituting gives $\frac{80}{16} = \frac{V}{9}$. Solving for V gives 45 liters per minute. **Answer is A.**

23. Multiplying equation 1 by 2, and equation 2 by –3, and then adding together the two resulting equations gives -11y + 22z = 0. Solving for y gives y = 2z. In the meantime, multiplying equation 3 by –2 and adding it to equation 2 gives
–y – 12z = -14. Then substituting 2z for y, yields the result z = 1. Subsequently, one can easily find that y = 2, and x = -1. **Answer is A.**

24. Using the definition of absolute value, two equations are possible: 18 = 4 + 2x or 18 = 4 – 2x. Solving for x gives x = 7 or x = -7. **Answer is C.**

25. Rewriting the inequality gives x^2 – 5x + 6 > 0. Factoring gives (x – 2)(x – 3) > 0. The two cut-off points on the numberline are now at x = 2 and x = 3. Choosing a random number in each of the three parts of the numberline, we test them to see if they produce a true statement. If x = 0 or x = 4, (x-2)(x-3)>0 is true. If x = 2.5, (x-2)(x-3)>0 is false. Therefore the solution set is all numbers smaller than 2 or greater than 3.
Answer is D.

MATHEMATICS HIGH SCHOOL

TEACHER CERTIFICATION EXAM

26. Possible rational roots of the equation $0 = x^3 + x^2 - 14x - 24$ are all the positive and negative factors of 24. By substituting into the equation, we find that -2 is a root, and therefore that $x+2$ is a factor. By performing the long division $(x^3 + x^2 - 14x - 24)/(x+2)$, we can find that another factor of the original equation is $x^2 - x - 12$ or $(x-4)(x+3)$. Therefore the zeros of the original function are -2, -3, and 4. **Answer is D.**

27. Getting the bases the same gives us $3^{\frac{1}{2}} 3^{\frac{2}{3}}$. Adding exponents gives $3^{\frac{7}{6}}$. Then some additional manipulation of exponents produces $3^{\frac{7}{6}} = 3^{\frac{14}{12}} = \left(3^2\right)^{\frac{7}{12}} = 9^{\frac{7}{12}}$. **Answer is B.**

28. Simplifying radicals gives $\sqrt{27} + \sqrt{75} = 3\sqrt{3} + 5\sqrt{3} = 8\sqrt{3}$. **Answer is A.**

29. Multiplying numerator and denominator by the conjugate gives
$\dfrac{10}{1+3i} \times \dfrac{1-3i}{1-3i} = \dfrac{10(1-3i)}{1-9i^2} = \dfrac{10(1-3i)}{1-9(-1)} = \dfrac{10(1-3i)}{10} = 1-3i$. **Answer is D.**

30. To find the 100th term: $t_{100} = -6 + 99(4) = 390$. To find the sum of the first 100 terms: $S = \dfrac{100}{2}(-6+390) = 19200$. **Answer is A.**

31. There are 3 slots to fill. There are 3 choices for the first, 7 for the second, and 6 for the third. Therefore, the total number of choices is $3(7)(6) = 126$. **Answer is A.**

32. The set-up for finding the seventh term is $\dfrac{8(7)(6)(5)(4)(3)}{6(5)(4)(3)(2)(1)}(2a)^{8-6} b^6$ which gives $28(4a^2 b^6)$ or $112a^2 b^6$. **Answer is C.**

33. By definition, parallel lines are coplanar lines without any common points. **Answer is B.**

34. A set of n objects has 2^n subsets. Therefore, here we have $2^4 = 16$ subsets. These subsets include four which each have 1 element only, six which each have 2 elements, four which each have 3 elements, plus the original set, and the empty set. **Answer is B.**

35. Formula for finding the measure of each interior angle of a regular polygon with n sides is $\dfrac{(n-2)180}{n}$. For n=10, we get $\dfrac{8(180)}{10} = 144$. **Answer is C.**

MATHEMATICS HIGH SCHOOL

TEACHER CERTIFICATION EXAM

36. Draw a right triangle with legs of 6 and 8. Find the hypotenuse using the Pythagorean Theorem. $6^2 + 8^2 = c^2$. Therefore, c = 10 miles. **Answer is B.**

37. The formula relating the measure of angle K and the two arcs it intercepts is $m\angle K = \frac{1}{2}(mPS - mAD)$. Substituting the known values, we get $10 = \frac{1}{2}(40 - mAD)$. Solving for mAD gives an answer of 20 degrees. **Answer is B.**

38. Given a point on a line, place the compass point there and draw two arcs intersecting the line in two points, one on either side of the given point. Then using any radius larger than half the new segment produced, and with the pointer at each end of the new segment, draw arcs which intersect above the line. Connect this new point with the given point. **Answer is D.**

39. By definition this describes the procedure of an indirect proof. **Answer is C.**

40. Since side AK is common to both triangles, the triangles can be proved congruent by using the Side-Angle-Side Postulate. **Answer is C.**

41. In an ordinary parallelogram, the diagonals are not perpendicular or equal in length. In a rectangle, the diagonals are not necessarily perpendicular. In a rhombus, the diagonals are not equal in length. In a square, the diagonals are both perpendicular and congruent. **Answer is C.**

42. The most one can say with certainty is that the median (segment drawn to the midpoint of the opposite side) and the altitude (segment drawn perpendicular to the opposite side) of a triangle <u>may</u> coincide, but they more often do not. In an isosceles triangle, the median and the altitude to the <u>base</u> are the same segment. **Answer is A.**

43. In the mid-nineteenth century, Reimann and other mathematicians developed elliptic geometry. **Answer is B.**

44. Let's assume the base of the rectangular solid (box) is 3 by 4, and the height is 5. Then the surface area of the top and bottom together is 2(12) = 24. The sum of the areas of the front and back are 2(15) = 30, while the sum of the areas of the sides are 2(20)=40. The total surface area is therefore 94 square feet. **Answer is C.**

45. Find the area of the garden and then subtract the area of the fountain: $30(60) - \pi(5)^2$ or approximately 1721 square meters. **Answer is A.**

46. To find the area of the shaded region, find the area of triangle ABC and then subtract the area of triangle DBE. The area of triangle ABC is .5(6x)(y) = 3xy. The area of triangle DBE is .5(2x)(y) = xy. The difference is 2xy. **Answer is B.**

MATHEMATICS HIGH SCHOOL

47. Find the mean: 300/8 = 37.5. Then, using the formula for standard deviation, we get $\sqrt{\dfrac{2(37.5-37)^2 + 2(37.5-38)^2 + (37.5-35)^2 + (37.5-40)^2 + (37.5-36)^2 + (37.5-39)^2}{8}}$ which has a value of 1.5. **Answer is B.**

48. To find the determinant of a matrix without the use of a graphing calculator, repeat the first two columns as shown,

```
2    1   -1    2    1
4   -1    4    4   -1
0   -3    2    0   -3
```

Starting with the top left-most entry, 2, multiply the three numbers in the diagonal going down to the right: 2(-1)(2)=-4. Do the same starting with 1: 1(4)(0)=0. And starting with –1: -1(4)(-3) = 12. Adding these three numbers, we get 8. Repeat the same process starting with the top right-most entry, 1. That is, multiply the three numbers in the diagonal going down to the left: 1(4)(2) = 8. Do the same starting with 2: 2(4)(-3) = -24 and starting with –1: -1(-1)(0) = 0. Add these together to get -16. To find the determinant, subtract the second result from the first: 8-(-16)=24. **Answer is C.**

49. Using the Pythagorean Identity, we know $\sin^2 x + \cos^2 x = 1$. Thus $1 - \sin^2 x = \cos^2 x$, which by definition is equal to $1/\sec^2 x$. **Answer is D.**

50. Tan A = 9/12=.75 and $\tan^{-1}.75$ = 37 degrees. Since angle B is complementary to angle A, the measure of angle B is therefore 53 degrees. **Answer is D.**

51. The composite function $g(f(x)) = (3x-2)^2 = 9x^2 - 12x + 4$. **Answer is C.**

52. Given the polar point $(r,\theta) = (5, 60)$, we can find the rectangular coordinates this way: $(x,y) = (r\cos\theta, r\sin\theta) = (5\cos 60, 5\sin 60) = (2.5, 4.33)$. **Answer is C.**

53. Using the Pythagorean Theorem, we get v = $\sqrt{36+25} = \sqrt{61}$. **Answer is B.**

54. The line x = 5 is a vertical line passing through (5,0) on the Cartesian plane. By observation the distance along the horizontal line from the point (-2,7) to the line x=5 is 7 units. **Answer is D.**

TEACHER CERTIFICATION EXAM

55. The formula for finding the midpoint (a,b) of a segment passing through the points $(x_1, y_1) and (x_2, y_2) is (a,b) = (\frac{x_1+x_2}{2}, \frac{y_1+y_2}{2})$. Setting up the corresponding equations from this information gives us $x = \frac{-4+2}{2}, and\, 1 = \frac{y-3}{2}$. Solving for x and y gives x = -1 and y = 5. **Answer is A**.

56. Dividing by 36, we get $\frac{x^2}{36} + \frac{y^2}{4} = 1$, which tells us that the ellipse intersects the x-axis at 6 and –6, and therefore the length of the major axis is 12. (The ellipse intersects the y-axis at 2 and –2). **Answer is C**.

57. With a diameter going from (0,7) to (0,3), the diameter of the circle must be 4, the radius must be 2, and the center of the circle must be at (0,5). Using the standard form for the equation of a circle, we get $(x-0)^2 + (y-5)^2 = 2^2$. Expanding, we get $x^2 + y^2 - 10y + 21 = 0$. **Answer is B**.

58. By definition of log form and exponential form, $\log_x k = m$ corresponds to $x^m = k$. **Answer is A**.

59. Use the Power Rule for polynomial differentiation: if $y = ax^n$, then $y' = nax^{n-1}$. **Answer is D**.

60. Use the Exponential Rule for derivatives of functions of e: if $y = ae^{f(x)}$, then $y' = f'(x)ae^{f(x)}$. **Answer is A**.

61. To find the slope of the tangent line, find the derivative, and then evaluate it at $x = \frac{\pi}{2}$. y' = 3x(-sinx)+3cosx. At the given value of x, $y' = 3(\frac{\pi}{2})(-\sin\frac{\pi}{2}) + 3\cos\frac{\pi}{2} = \frac{-3\pi}{2}$. **Answer is A**.

62. To find the slope of the tangent line, find the derivative, and then evaluate it at x=1. y'=6x-5=6(1)-5=1. Then using point-slope form of the equation of a line, we get y-3=1(x-1) or y = x+2. **Answer is C**.

63. To find critical points, take the derivative, set it equal to 0, and solve for x. f'(x) = $3x^2$ + 2x = x(3x+2)=0. CP at x=0 and x=-2/3. Neither of these CP is on the interval from x=1 to x=3. Testing the endpoints: at x=1, y=6 and at x=3, y=38. Since the derivative is positive for all values of x from x=1 to x=3, the curve is increasing on the entire interval. **Answer is B**.

TEACHER CERTIFICATION EXAM

64. Find CP at x=-.75 as done in #63. Since the CP is not in the interval from x=0 to x=3, just find the values of the functions at the endpoints. When x=0, y=0, and when x=3, y = 27. Therefore 27 is the absolute maximum on the given interval. **Answer is D.**

65. Use the rule for polynomial integration: given ax^n, the antiderivative is $\dfrac{ax^{n+1}}{n+1}$. **Answer is A.**

66. Use the rule for integration of functions of e: $\int e^x dx = e^x + C$. **Answer is D.**

67. Recall that the derivative of the velocity function is the acceleration function. In reverse, the integral of the acceleration function is the velocity function. Therefore, if a=6, then v=6t+C. Given that at t=0, v=15, we get v = 6t+15. At t=10, v=60+15=75m/s. **Answer is C.**

68. Recall that the derivative of the distance function is the velocity function. In reverse, the integral of the velocity function is the distance function. To find the time needed for the body to come to a stop when v=0, solve for t: v = 16 – t² = 0. Result: t = 4 seconds. The distance function is s = 16t - $\dfrac{t^3}{3}$. At t=4, s= 64 – 64/3 or approximately 43 units. **Answer is B.**

69. Integrate as described in #65. **Answer is B.**

70. Use the fundamental theorem of calculus to find the definite integral: given a continuous function f on an interval [a,b], then $\int_a^b f(x)dx = F(b) - F(a)$, where F is an antiderivative of f.

$\int_0^2 (x^2 + x - 1)dx = (\dfrac{x^3}{3} + \dfrac{x^2}{2} - x)$ Evaluate the expression at x=2, at x=0, and then subtract to get 8/3 + 4/2 – 2-0 = 8/3. **Answer is B.**

71. To find the area set up the definite integral: $\int_3^6 (x^2 + 4)dx = (\dfrac{x^3}{3} + 4x)$. Evaluate the expression at x=6, at x=3, and then subtract to get (72+24)-(9+12)=75. **Answer is A.**

MATHEMATICS HIGH SCHOOL

TEACHER CERTIFICATION EXAM

72. The errors are in the following: -3+7=-4 and –3 + -8 = 11, where the student seems to be using the rules for signs when multiplying, instead of the rules for signs when adding. **Answer is C**.

73. Although all of the answers have some merit, **answer B seems the best.**

74. Since the second equation is already solved for y, it would be easiest to use the substitution method. **Answer is C.**

75. The steps listed in answer D would look like this for the given example:
If $|x+1| \leq 6$, then $-6 \leq x+1 \leq 6$, which means $-7 \leq x \leq 5$. Then the inequality would be graphed on a numberline and would show that the solution set is all real numbers between –7 and 5, including –7 and 5. **Answer is D.**

76. There is no need for calculators when justifying statements in a geometric proof. **Answer is C.**

77. By observation the **Answer is B**.

78. All of the answers have some merit, **but C and A are the best answers.**

79. All of the answers have some merit depending on the software available, but **C is the best answer.**

80. By observation of the examples given, $a \not\subset b = a^2 - b$. Therefore, $5 \not\subset 4 = 25 - 4 = 21$. **Answer is D**.

Good luck to all of you. May you have many happy years teaching.
The end.

XAM Publishing, Inc. neither endorses or represents the following website but offers it your consideration as a resource.

www. Interactivemathtutor.com

TEACHER CERTIFICATION EXAM

"Mrs. Hammond, I'd know you anywhere from little Billy's portrait of you."

MATHEMATICS HIGH SCHOOL

TEACHER CERTIFICATION EXAM

"Are we there yet?"

MATHEMATICS HIGH SCHOOL